BUILDING, PLANNING AND DEVELOPMENT

ERNEST H. GREEN, F.R.I.C.S., F.I.A.S., F.R.V.A.

Consultant and External Tutor,
College of Estate Management,
one time Principal Lecturer and Deputy Head
of Department of Professional Studies,
West London College

First published 1981 by
THE MACMILLAN PRESS LTD
London and Basingstoke
Associated companies in Delhi Dublin
Hong Kong Johannesburg Lagos Melbourne
New York Singapore and Tokyo

Printed in Hong Kong

British Library Cataloguing in Publication Data

Green, Ernest H
    Building planning and development. – (Macmillan
    building and surveying series).
    1. Real estate development – England
    I. Title
    333.3'8   HD608

    ISBN 0–333–19788–7
    ISBN 0–333–19789–5 Pbk

BUILDING, PLANNING AND DEVELOPMENT

Macmillan Building and Surveying Series

Advanced Building Quantities   Ivor H. Seeley and R. Winfield

Building Economics   Ivor H. Seeley

Building Maintenance   Ivor H. Seeley

Building Technology   Ivor H. Seeley

Building Quantities Explained   Ivor H. Seeley

Computers in Quantity Surveying   R. J. Alvey

Introduction to Valuation   D. Richmond

National and Local Taxation   Michael Rayner

The British Construction Industry: an introduction   Dennis F. Dolan

Urban Land Economics   P. N. Balchin and J. L. Kieve

Related Macmillan Titles

Administrative Law for the Construction Industry   J. R. Lewis

Law for the Construction Industry   J. R. Lewis

CONTENTS

PREFACE

This textbook is intended primarily for students who are in the final stages of their studies for a degree or professional qualification in one of the landed professions. The author makes no attempt to follow the exact syllabus of any particular body because they are often subject to change, but hopes that the book will be of value to many.

A large development demands many skills and professional specialisations and no one person could be equipped to deal with all the stages in the development process. It is important nevertheless for all those taking part to have at least an understanding of the contributions made by others.

Each chapter deals with a different aspect of the development process and each, if studied in isolation, could be shown to be incomplete—but that is inevitable with a book of this kind.

It will not teach any particular member of a development team his job but it should enable him to see a development scheme in a wider perspective and this in turn should make each individual contribution more valuable.

This book is written at a time when little development is taking place, at least by comparison with the boom period of the late 1960s and early 1970s. Building costs have virtually doubled in the past few years and interest rates have been high. This has made it more costly to borrow money and it has also meant that the return in the form of rent on new development has to be proportionately higher than was generally acceptable a few years ago.

The author is conscious of this situation and therefore appreciates that some of the material used in dealing with the financial aspects of development may not be strictly applicable to conditions as they exist today, but in order to produce a balanced textbook it is necessary to assume reasonably stable conditions, and it is hoped that these will prevail when the book is in circulation.

ERNEST GREEN

ACKNOWLEDGEMENTS

I am grateful to Ian Hamilton-Penney, M.S.A.A.T., A.I.P.D., for most of the illustrations used in the book, and to Ken Blount, F.I.A.S., M.I.L.G.A., A.M.B.I.M., Chief Building Surveyor for Wolverhampton Borough Council, for his guidance on building control.

I also wish formally to acknowledge the help given to me by: Town and City Properties Ltd on my visits to the Arndale Centre at Luton; and The Bradford Property Trust, for plans and other information on the development of a new village at Martlesham Heath near Ipswich.

E.H.G.

1 INTRODUCTION— AN HISTORICAL APPROACH

Building development and planning are activities that are very much concerned with the everyday life of ordinary people. Factors such as where people live, where they work and where they spend the reward for that work all create a demand for some form of development. Indeed most human activities create a demand for buildings or building work of some kind. The ability to meet this demand depends on the extent of our natural resources and this constraint, taken with the complexities of modern living, has made planning essential and some form of statutory control over development at least desirable.

It has often been said that history has little relevance to planning to meet the basic human needs of the future, and there is no doubt that it would be a mistake to be a slave to the past. On the other hand a student would be ill advised to ignore completely the good work that has been done in the past. It may also help to examine some of the failures, if only to learn what to avoid. The United Kingdom has some fine buildings and many interesting towns and villages. Developments like the Georgian terraces in Bath are world famous, and there must be something to be learned from a study of such work. A number of industrial villages, such as those built by Titus Salt, Robert Owen, George Cadbury and William H. Lever, still serve a useful purpose today and to a student of planning and development these are equally worthy of study. In the nineteenth century a number of schemes that were put forward to create completely new communities did not get beyond the planning stages but some of the ideas have been found useful by today's planners.

Robert Owen was best known for the growth and development of New Lanark. This is a mill village, which was started in 1784 by Dale and Arkwright, and taken over by Owen in about 1800 when he married Dale's daughter and became manager. In little over a quarter of a century he established a model industrial village. The site layout was not ideal since the building work was not based on a master plan but was carried out on an *ad hoc* basis. Using the lessons he learned at New Lanark (figure 1.1) he was convinced that the only way to overcome the problems of the poor was to set up what he called villages of co-operation where people could live and work under improved conditions and thus be saved from the misery and squalor which were the lot of most of the workers in rapidly expanding industry. The village (figure 1.2) was to have a population of up to 1200 and to consist of an open-plan square with gardens, and three principal buildings—a public kitchen and schools at two levels. The open square was enclosed by terraces of buildings: on three sides were terraces of workers' houses, with a centre block reserved for administrators, teachers and doctors, who were given more generous accommodation. On the fourth side the terrace contained dormitories for children over three, a hostel for visitors at one end and an infirmary

1

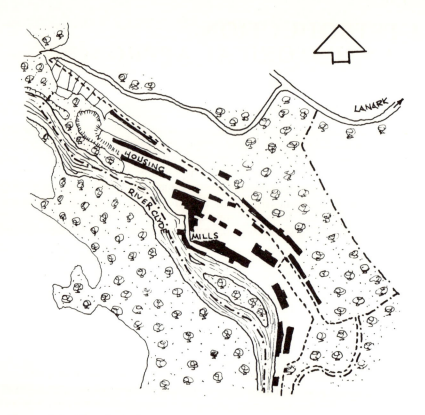

Figure 1.1   New Lanark

at the other. On the outside of the square were private gardens bounded by main roads; outside there were wash houses, a slaughter house and 'two farming establishments with cornmill, malting and brewing appendages'. All this was to be set in about 1200 acres of mainly agricultural land. Owen even produced a schedule of expenses showing the estimated cost for the whole scheme to be £96 000.

The supporters of Robert Owen could rightly claim that he left behind at least one basic planning principle, 'that it was just as important to plan for the workers as for the machines they operated and that it made sound economic sense to provide them with a more comfortable home in a well-planned environment'.

A report to the Health of Towns Commission in 1848 drew public attention to the appalling conditions in new industrial suburbs. Titus Salt was Mayor of Bradford at about the time of this report and this fact, coupled with the squalid conditions he encountered during his term of office, caused him to abandon his plan to retire. Instead he planned to build a new mill and an associated industrial village on a site some miles out of the town. Salt's village was called Saltaire and now forms part of Bradford; the village is well

2

preserved and almost intact. The density of the residential development is high
by modern standards but the architects produced a scheme which met all the
needs of such a community. The most important fact about this effort was that
it was planned as one unit and although the work was spread over a number of
years it was built to the original plan, which is reproduced in figure 1.3.

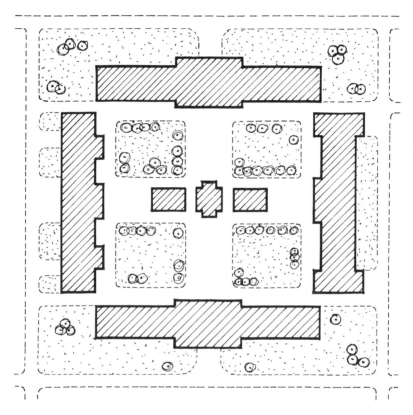

*Figure 1.2   Diagrammatic representation of Robert Owen's village of
co-operation*

Salt was both a practical man and a man of wealth and was therefore able to
put into practice his idea of reform. Many people were putting forward ideas
of Utopia which were quite impracticable, and little was done to try to bring
them into being, but they probably had some influence on George Cadbury,
who in 1879 started building a new factory about 4 miles to the south-east of
Birmingham. Houses were built for the workers and by the turn of the century
Bournville was a well planned and thriving community occupying over 120
acres of land. The foundation of a Trust to own and manage the houses was
sponsored by the family firm and remained under its guidance but not under

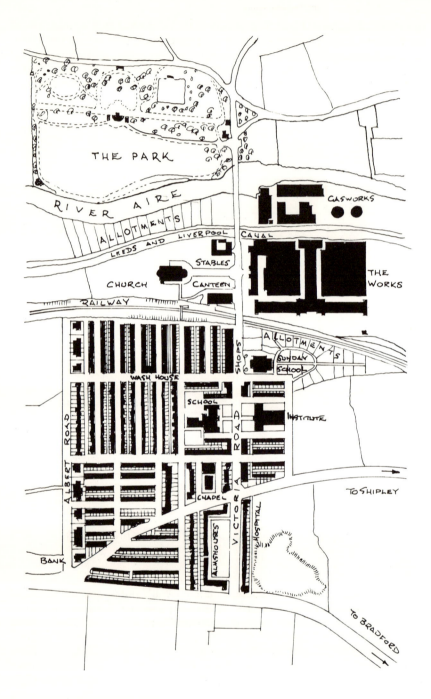

Figure 1.3 Plan of Saltaire

4

its control. The estate has now grown to over 900 acres; figure 1.4 is a plan of the estate as it was in 1898. It is still a pleasant place in which to live and is more spacious than Saltaire. This is due to the advanced thinking of George Cadbury, who insisted that 'the houses must not occupy more than one-quarter of their sites and at least one-tenth of the land in addition to roads and gardens should be reserved for park and recreation ground'.

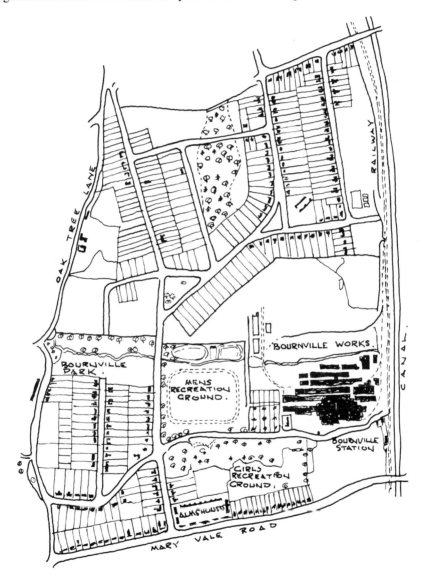

Figure 1.4   Plan of Bournville

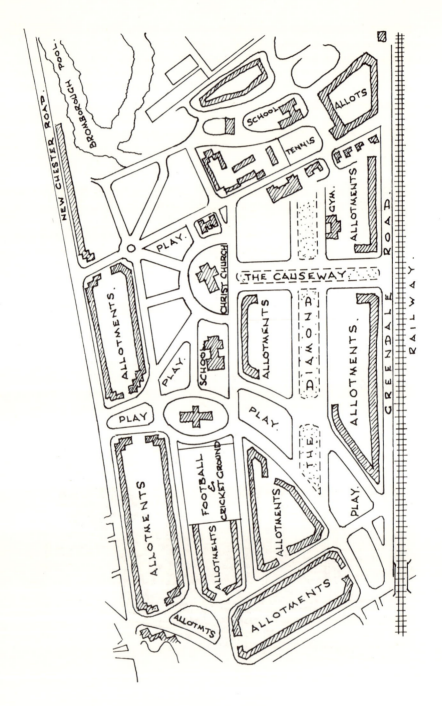

Figure 1.5 Plan of the village of Port Sunlight

The Bournville scheme has much of the paternalism of both New Lanark and Saltaire, but the residential part of the development was hived off into the ownership of what is now a housing trust and this gave the residents an independence not enjoyed by those of either New Lanark or Saltaire. William Heskith Lever built Port Sunlight and adopted many of the principles that were followed by Cadbury. In 1888 he decided to move his soap factory from Warrington to a new site on the River Mersey near Birkenhead. He purchased 52 acres of land on which to establish an industrial village. He drew up his own plan, which is reproduced in figure 1.5, showing a more compact layout than that of Bournville (figure 1.4). A particular characteristic of this scheme was that the houses were sited around open plots of land devoted to gardens and allotments. Both the developments at Bournville and at Port Sunlight have continued into the twentieth century and are still very much part of the urban pattern in their localities.

During the same period another form of mass development was taking place in London and other large cities, and also in the spa towns, whose waters attracted people to drink or bathe to benefit from their curative powers. This development started in what is called the Georgian period. Much has been written about this important period of architecture but its real value to a student of development is that the estates were planned as one scheme and although the actual construction may have been carried out by different builders, the road pattern and the siting of the central gardens were subject to a master plan. The square was the form most frequently used but circuses, crescents and ovals were also used. The siting of the square could be varied, as will be seen from the treatment of Portman Square and Manchester Square shown in figure 1.6. The use of the crescent is well demonstrated in the

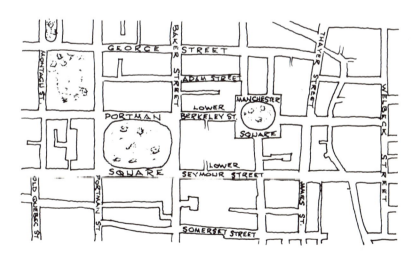

Figure 1.6   Squares in London's West End

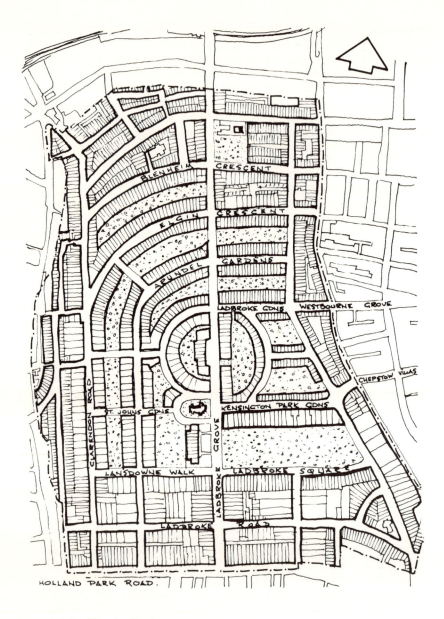

Figure 1.7   The Ladbroke estate

Ladbroke estate (figure 1.7), which was built some time later. Thomas Cubitt was probably the forerunner of the developer as we know him today. His best-known work began in the 1820s when the Duke of Bedford decided to continue with the development of his Bloomsbury estate. Cubitt built

Tavistock Square, Woburn Place, part of Gordon Square and a number of streets in the vicinity. He then went on to develop a large area of the Grosvenor estate, which is now known as Belgravia. No provision was made for working-class tenants except for domestic servants, the coachman often living in the mews. Cubitt provided a church and schools but no social buildings, except pubs, which he usually built first to provide refreshments for his large work-force of over 1000 at one time.

Towards the end of this period of extensive growth in quality housing, some efforts were being made in London to improve the standard of houses for the working classes. From 1840 onwards a number of associations were formed to improve the dwellings of the 'industrious working classes' with the object of 'supplying the labouring man with an increase of all the comforts and conveniences of life with full compensation to the capitalist'. There was a limit of about 5 per cent return on capital invested.

In 1860 Sidney Waterlow formed a company and a little later George Peabody and others provided funds for building. Much of this development was in the form of tenements, many of which are still in use today. Most are in the process of being improved and brought up to an amenity standard as high as is practicable with such structures. The buildings are a little stark and austere but provide accommodation for a large number of people, and because of the housing shortage in the capital it would be difficult to justify their demolition.

At the end of the nineteenth century a new concept in community planning and development started with an idea put forward by a shorthand writer working in the Houses of Parliament. He was Ebenezer Howard, who set off a chain of events which still has considerable influence on modern thinking. He began by writing a book which was an attempt to plot a peaceful path to real reform. The book was first published in 1898 as *Tomorrow* but in later editions the name was changed to *Garden Cities of Tomorrow*. It put forward a new idea, that of creating a township in the country which was to have the advantages of both town and country. Howard visualised the garden city as one complete social unit, with its population controlled and its growth limited, the overspill from one garden city forming the beginnings of another. To be sure that there would be no encroachment by other developments, 6000 acres were to be acquired but only 1000 used for development. The remainder were to be in permanent agricultural use to ensure a pleasant environment and to grow food. This was one of the early suggestions for what is now known in planning as 'green belt' policy.

Most of Howard's diagrammatic plans took a circular form (a segment of one is shown in figure 1.8) but he said that when his ideas were adopted this would be much departed from. This is borne out by the plan for the first garden city at Letchworth (figure 1.9). The site selected was bisected by a railway line, which formed a physical barrier at many points, as can be seen from the plan in figure 1.10. In addition, neither of the two architects Raymond Unwin or Barry Parker was attracted to mechanically produced design patterns. Their dislike for uniformity and their desire to secure variety may give some people an impression of uncontrolled development which fortunately has been counteracted by the planting of many trees and the

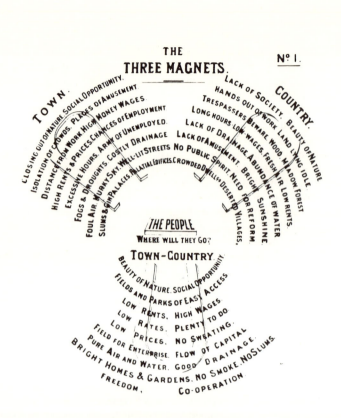

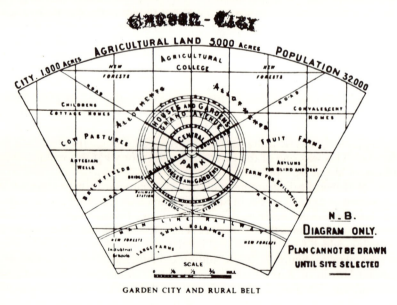

Figure 1.8   Howard's Three Magnets

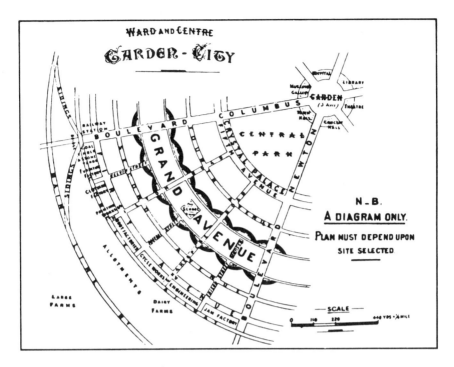

Figure 1.9   Garden city diagram

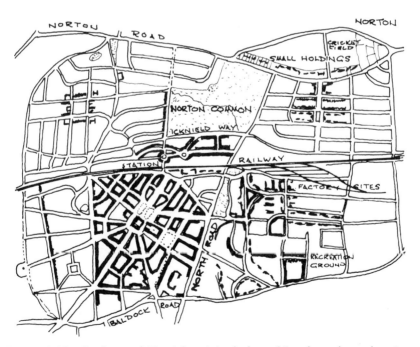

Figure 1.10   Parker and Unwin's original plan of Letchworth garden city

11

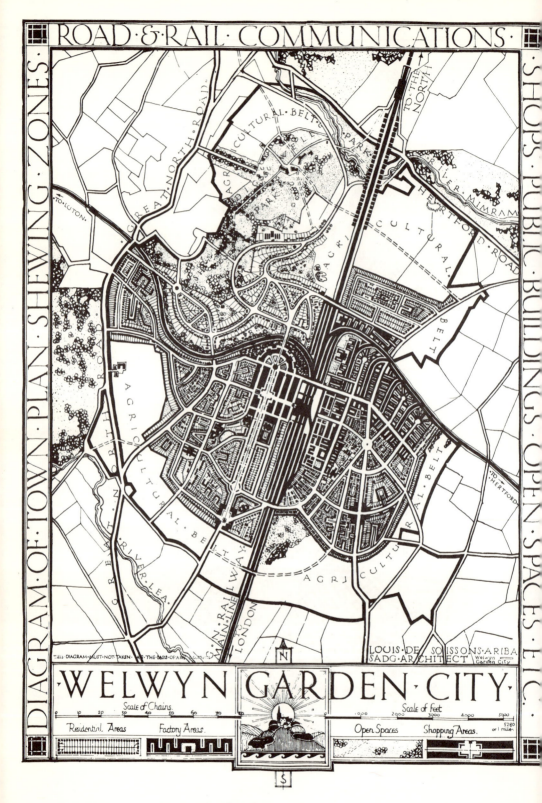

Figure 1.11   Welwyn Garden City

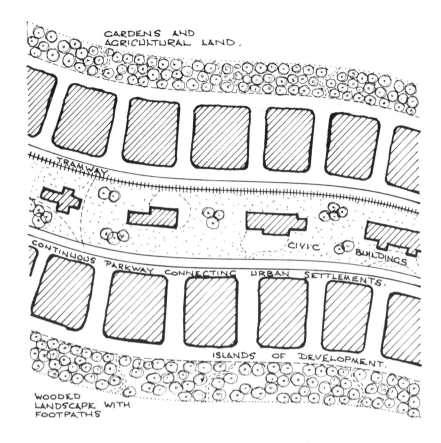

Figure 1.12   The linear city of Don Arturo Soria y Mata

provision of wide grass verges. Howard's second garden city at Welwyn is considered the better of the two, the architects having learned from their earlier experience. Welwyn is illustrated in figure 1.11.

An interesting idea for a new town was put forward in 1882 by Arturo Soria y Mata, a Spanish architect. It was not entirely new since it was based on controlled ribbon development. His plan, which is shown in diagrammatic form in figure 1.12 is for a linear town of population 30 000. The spine was to be a road or parkway 300 m wide, with a tram track with generous green verges and central reservation on which were to be sited public buildings and utility services. The rectangles of development on either side of the road were to be limited to a depth of 200 m and were to be bounded by strips of woodland opening on to natural countryside. He was able to try out his idea in a new development but it made poor progress and was eventually abandoned.

In almost exact contrast to the concept of the linear town was one of the many ideas put forward by Le Corbusier. This was for one large self-contained block with dwellings, shops and services all under one roof. This idea, probably inspired by his earlier writing, can be seen in Dolphin Square in London and in the large council developments in Sheffield known as Hyde Park and Park Hill.

It is convenient to treat the Second World War as a milestone in planning and development in the United Kingdom. No permanent development, other than that directly concerned with the war effort, took place over the war period and for some time afterwards. The present era of building development and planning can be said to have started in July 1948, which was the operative date for many of the provisions of the Town and Country Planning Act, 1947, and therefore the beginning of planning control as we know it today.

2 PARTIES TO A DEVELOPMENT

Throughout this book the word 'developer' is used to denote the one or many persons or firms that are engaged in a development scheme. On the one hand the developer could be a major property company occupied in redeveloping a town centre or, on the other hand, a small local firm building a dozen houses.

This very general term has been adopted for reasons of simplicity and it now becomes necessary to take a careful look at the organisations and professions which are likely to be concerned with development. If we consider the part they play in the scheme, they fall conveniently into three different but interrelated roles

(1) development agencies
(2) professional and technical advisers
(3) the contractor.

Each of these roles will be examined in turn.

DEVELOPMENT AGENCIES

This is a convenient name for the persons or companies taking financial responsibility for the enterprise. The financial aspect may be the only thing that the agencies have in common since they differ in all other respects, each having its own objectives, decision-making procedures and financial structure. Their reasons for being in the development business are all different and can cover a wide field. At one end is the entrepreneur with his profit motive and at the other end the local authority with its moral and social obligations, and to some extent its statutory duty to provide housing. Although they have entirely different motives these two widely differing development agencies do sometimes get together in town centre renewal; this is discussed in chapter 11.

The objectives of the development agency and its reasons for being engaged in this field of activity will influence many aspects in the planning of a development scheme. Although a detailed examination of the many variations that exist is not necessary for a book of this kind, it is essential for the student to have a knowledge of the basic differences in so far as they influence the design requirements and the financial aspects of the scheme.

Central Government

Government departments are often cast in the role of developer in such varied schemes as the construction of government buildings or the provision of married quarters for Service personnel. Most of this work is handled by the Property Services Agency, which grew out of the Office of Works, and which was formed for this purpose. Development by the PSA is not usually of a

15

speculative nature but takes the form of schemes to meet specific requirements.

Local Government

Local authorities are constantly concerned in many forms of development but in most cases their greatest involvement is in the field of slum clearance and residential development. In order to qualify for grants and subsidies and also to obtain loan sanction, new development has to meet certain design criteria. These criteria are at present based on recommendations contained in the Parker Morris Report.[1] Limitations are also imposed on building costs. These are based on a cost yardstick published from time to time by central government and special application has to be made for these standards to be varied. More information on this financial restraint is given in chapter 6. Large-scale new development can take place when a large local authority gets together with a small one to co-operate in an expanding town scheme under the Town Development Act, 1952. Most of the design criteria and cost controls also apply to this form of development but in many ways the end product is similar to that produced in the new towns.

Activity in this field has been running down for the past three or four years, and some schemes have already been wound up substantially short of their original target figures. This action is due to a number of different factors, not least to the deterioration in inner city areas caused by too many industries and their workers moving out to new or expanding towns. The change of emphasis from development in expanding towns to redevelopment in inner city areas will mean that many more city councils are likely to be involved in industrial and commercial schemes than hitherto.

New Town Corporations

As a development agency the corporation of a new town is involved in all forms of development and has wide powers, including the power of compulsory purchase and the management and disposal of land and property. Compensation is based on the value of land prior to the making of the order designating the area of land for the new town and not on the proposed use of the land in the new development. This means that land can be allocated more generously than would be possible with most development agencies. The corporation has the best of both worlds—it can build to sell, rent or let on long lease and it can lease or sell land for private development in the area designated. The corporation is not required to make a profit but in the long term the whole project must be financially viable.

Public Utilities and Nationalised Industries

These are many and varied, with different constitutions and different financial structures. As development agencies they usually work on specific instructions and they can therefore be treated as quasi-government departments.

Private Developers

Notwithstanding the growth in State ownership, development schemes are still being carried out in residential, commercial and industrial fields and vast sums of money are invested in completed developments. (See the note in the preface concerning the economic conditions at the time of writing this book.) The test of viability in the case of a private development must be either immediate profit or long-term investment with growth. Without one of these the necessary funds could not be raised to finance the scheme. The completed development has to compete in the market with other similar schemes and it is not cushioned against drastic changes in the economic climate as are municipal schemes.

Housing Societies and Associations

These are generally active in residential development and they usually build houses to rent. Old-established charitable trusts fall into this category and as development agencies they occupy a special position. Provided they are properly constituted and are approved by the Housing Trust they are able to seek loans and other financial help from local authorities, who have the power to acquire development land for them.

A professional adviser receiving an instruction to plan a development will need to know his clients and to which category of development agency they belong. This will materially affect the design considerations and many other aspects of the development scheme. This important point is further brought out in later chapters, particularly in chapter 8, which deals with residential development.

PROFESSIONAL AND TECHNICAL ADVISERS

A development agency will need both professional and technical advice. This may come from outside consultants or from inside the organisation. The size of the scheme and the type of development will dictate the number of advisers needed and the specialisations involved. A large comprehensive development scheme will require a team of experts, with each expert contributing his own particular specialist knowledge and expertise to an overall plan, but these specialisations should not be looked on as being in watertight compartments since the work overlaps in most of the professions. With a smaller development a firm of general practice surveyors may handle all aspects of the scheme except the actual work of construction.

The Development Team

The author makes no attempt to place the advisers in any order of importance, neither does he suggest any limits of demarcation between the different functions.

The Valuer   Early in any scheme a practising valuer will be required to give advice and expertise concerning the value of the site and also to act in the negotiations for its purchase.

The Architect   The design of buildings is the main function of the architect. He does this by bringing together scientific principles and technical information. By using his imagination he then creates a project capable of meeting his client's specific requirements economically, with efficiency and within the constraints imposed.

The Land Economist   Research into the demand for and the supply of the type of development proposed is the work of a land economist. The conduct of surveys and the analysis and interpretation of these surveys will contribute substantially to the assessment of the viability of any proposed scheme.

The Quantity Surveyor   Costing is obviously one of the most important aspects of any development scheme and this is the specialised field of the quantity surveyor. His main responsibility is still the preparation of bills of quantities but he may also be concerned with cost planning, cost analysis and valuations for interim payments.

Engineering Consultants   These will be called in to advise on specific matters that are outside the scope and experience of other members of the team. In some types of development much of the design work may be done by a structural engineer.

The Landscape Architect   The adoption of open space and the duty to get the best out of that open space is the function of the landscape architect. The advice sought should cover not only the planting of trees and shrubs and the laying out of gardens, but also the treatment of the hard landscape, which includes paving and other hard surfaces.

The Accountant   Advice on the availability of finance demands a knowledge of the banking world, and the important function of control over cash flow calls for a knowledge of accountancy.

The Estate Agent   The disposal of the completed development by sale or on lease is best carried out by established agents, who should be experts in this specialised field. In a large-scale residential development it is common practice to set up a sales office on the site.

The Contractor   A knowledge of the relationship between the development agency and the professional and technical advisers is essential before any form of development is considered. It is also important to understand the contractual relationship between the contractor and any subcontractors employed.

   The contractor is usually selected by a tendering procedure and the choice depends on many factors, including the size and complexity of the proposed development. Objective tendering is the selection of the contractor who submits the lowest price in open competition. The pitfalls in this procedure may lead to the selection of a contractor who is quite incapable of doing the job. In subjective tendering a single contractor is nominated without any comparison being made with any other possible contenders for the contract. Because this procedure also has obvious pitfalls, most contracts are negotiated by a compromise procedure known as selective tendering. Here a number of possible contractors are selected, who are known to have financial, practical and staff resources adequate to carry out the scheme efficiently and on time. These contractors are given a copy of the brief and are invited to submit tenders, which are judged on a competitive basis. A contract is then drawn up between the development agency (thereafter known as the employer) and the contractor.

Subcontractors are often used to carry out specialised work, which may or may not be covered by patent. The contractor may also put out work to subcontractors to reduce his own risk, particularly for work in certain specialised trades. Although a subcontractor may be nominated by the architect, the contractual relationship is between the contractor and the subcontractor and no such relationship exists between the employer and the subcontractor.

REFERENCE

1. *Homes for Today and Tomorrow: Report of a subcommittee of the Central Housing Advisory Committee*, Chairman Sir Parker Morris (HMSO, 1961)

3 RECEIVING INSTRUCTIONS AND PRELIMINARY INVESTIGATIONS

THE INSTRUCTIONS

A good firm of general practice surveyors is likely to have the staff and experience to advise a client on almost any type of development project. This advice can take many forms but a frequent instance is that where a client is considering the purchase of a site for development. In other cases the client may own the site and be seeking advice on its best use. Occasionally instructions may be received to find a site suitable for a specific purpose or use. This did occur in the early stages of new town development but has become less frequent as government policy towards this form of development has changed. As a general rule, by the time a development scheme is being considered, the planning authority's views concerning land use will be known and any new development will be subject to its control.

Instructions can come in many different ways and it is useful for the person receiving the instructions to ask himself the question: 'Do I know exactly what I am required to do?' A valuable job may begin with a casual meeting with an old client, who says, 'It is about time I did something about developing that paddock near the village for housing. Look into it for me.' In such a case much more information will be needed and a proper client/adviser relationship established before any detailed work can be carried out.

At the other end of the scale instructions can come from a solicitor in writing and in considerable detail, but even in these circumstances some items of information may still be required before work can be carried out.

In both these cases and for all other ways in which instructions can be received, it is useful to have an itemised list against which an instruction can be checked. This document could well be of standard design, preprinted and used for all instructions of the kind that come into the office.

Fees can be the cause of misunderstanding between the professional adviser and his client and the extent of the instructions should therefore be fully understood and agreed by both sides. It is sometimes advisable to suggest to a client that the instruction is phased—the first phase in the case of a development project could well be up to and including the submission of an outline planning application. Because of the uncertainties that exist at the early stages of a development scheme it may be advisable to limit a client's expenditure, in the first instance, to that required to establish whether the scheme is approved in principle. If an outline consent is granted the client would then feel on firmer ground to authorise further and much greater expenditure on the next phase.

PRELIMINARY DOCUMENTATION

It is essential to have readily available the address of the subject site and the names and addresses of all those persons and organisations who may have to be contacted during the early stages of the project. The following will serve as a check list.

Location of Site

There should be no possible doubt concerning the exact location of the site and the record should contain

(1) the full postal address, including any street numbers, even if they are not generally used by the occupier
(2) a map reference if available.

As soon as the site is fully identified an Ordnance Survey sheet at a scale of 1/2500 should be used to define the site. This will be required in any case when the planning application is submitted.

The Client

The client may be a private individual, in which case it may be necessary for the surveyor to suggest a course of action to his client, particularly in the early stages of their relationship. If the client is a partnership or a corporate body it is important to know from whom instructions can be obtained and by whom decisions are to be taken. It is also important to know how soon a decision can be expected. Delay in decision-making may occur owing to the need to wait for a board meeting in the case of a company, or a committee or full council meeting in the case of a local authority.

The Client's Interest

The client may be the owner or lessee of the entire site or only part of the site, or he may be proposing to buy the site with or without planning permission. It is important to have an exact record of the client's interest in the land. This is likely to concern the surveyor when he wishes to have access to the land during his investigations. The information will also be required when a planning application is submitted since there are notices to be served and certificates to be completed as part of the process of making the application.

Other Interest

Notes at this stage can generally be limited to a record of

(1) the ownership of any land not owned by the client
(2) the leaseholders or tenants in possession.

A particular note should be taken of agricultural tenants, because one of the certificates referred to above concerns land being farmed. If any of the above have professional advisers their names and addresses should be recorded and a note made to say whether they can be contacted direct.

The Client's Agent

If the client is in the process of purchasing the land the name and address of his agent are required and possibly also an authority from the client giving the right of direct contact if necessary.

The client may not employ an agent to act in the purchase. He may do the negotiation himself or he may instruct the surveyor to act in the purchase of the site also.

The Vendor's Agent

Even if the surveyor is not also acting in the purchase of the land he will need to note the name and address of the vendor's agent, together with whatever other details are available, including the purchase price and the likely date of completion if agreed. Contact with the vendor's agent should normally be restricted to arrangements for rights of access before the purchase is completed. If a planning application is made before completion, the owner of the site at the time may also be the subject of a notice and certificate referred to earlier.

The Client's Solicitor

The extent to which the client's solicitor is involved in the early stages of a development scheme depends on many factors. In some situations he may take on the role of the client if, for example, the land is owned by a trust and the solicitor acts for that trust with full powers of attorney. A note should be made of the name of the firm and the name of the partner dealing with the matter, together with additional information about any particular authority held by the individual or firm.

Other Professional Advisers

The client may already have professional advisers—the most likely adviser is an architect. In such a situation it is vital that the client gives specific instructions in respect of what he looks for from each of his advisers. It is usual for the client to arrange the first meeting since it is important that a sound working relationship is established as early as possible.

Central and Local Government Departments

Matters connected with town and country planning at local level are now, in the main, dealt with by the district council. A note of the name of the council and the address of the planning department is necessary. It would also be of value to have the name of the planning officer and the name and qualifications of the assistant likely to be dealing with the matter of the application. If the proposals are likely to include office or industrial development, additional information will be required in this respect from the regional offices of central government departments. It may be necessary to have a note of the address of the local river authority and other bodies, such as the National Trust, and even the National Coal Board if the site is in an area of coal-bearing land.

Undertakings

Negotiations with the undertakings will usually come at a later stage in a development but a note of the names and address should be made early in the process. The local water authority may need to be consulted at an early stage in certain districts where development is restricted owing to a limitation on water supply or problems concerning sewage disposal and other drainage matters. Undertakings generally include

(1) the gas board
(2) the electricity board
(3) the water board
(4) GPO telephones
(5) local public transport.

PRELIMINARY INVESTIGATIONS AND ADVICE

A situation may occur where work has already been carried out on the project or on an earlier project on the same site, by other consultants. In such a situation it is essential to establish a correct relationship before detailed work is begun. If the instructions include work which in the first place was given to someone else, the surveyor should make certain of his exact position and if necessary consult his professional body about his ethical responsibilities in the matter. If he is still in doubt he may well be advised not to accept the instruction. Reference has already been made to a situation where an architect is already working for the client.

FEES AND EXPENSES

Most clients need to be informed at quite an early stage of the extent of their financial commitment in respect of fees and expenses. It may be difficult to estimate the amount of work to be done but as much detail as possible should be given to the client and this should include a reference to any scale of charges being used. This may be done by an exchange of letters or it may be desirable to draw up a suitable contract.

LEGAL MATTERS

Most of the legal enquiries will be the responsibility of the client's solicitor, but the surveyor will need to consult with him and in some cases make his own additional investigations in respect of the following.

Freehold

If the client's interest is freehold there will be a need to establish that he has, or can obtain, full vacant possession. If there are leasehold interests to be taken into account the following details should be ascertained

(1) the period of lease
(2) the rent or ground rent
(3) the position in respect of security of tenure
(4) the responsibilities on reversion.

Leasehold

If the client's interest is leasehold the investigation should cover all the points in (1) to (4) above.

Proposed Purchase

Details will be required of the terms of the purchase, the stage reached and in particular a note of any time limit which could be imposed by the vendor.

Restrictive Covenants

Covenants which go with the land and restrict its free use can sometimes hamper a development scheme to the extent that it becomes unworkable. Steps should be taken as early as possible to make a reference to the Lands Tribunal in an attempt to have such covenants removed or modified.

Easements and Other Rights

These can take the form of rights enjoyed by the client over adjoining land or rights over the site enjoyed by other persons or organisations. Those which are most likely to affect development are

(1) rights of light, and/or
(2) rights of way.

There may also be rights of entry on to the land for shooting or fishing. It might be possible to negotiate the termination of those rights which interfere with development.

Public utility undertakings may have wayleaves in respect of pipes and cables either above or below ground. The exact location of these will need to be plotted and recorded on a map at a very early stage in the planning process.

Boundaries

The exact boundaries should be established at an early stage together with, in particular, the client's responsibilities in respect of their maintenance and upkeep. The subject is dealt with in more detail in chapter 5.

LOCAL AUTHORITY ENQUIRIES

The local authority will be concerned in many aspects of a development scheme. It controls much of the building work from the point of view of planning and health and safety and will supply some of the services. Most of the information likely to be required at the early stages of a project will be obtainable at district council level.

Town and Country Planning

Early reference should be made to the structure plan, to the local plan and to any other 'plan' concerning the area in which the site is situated. The function and use of these plans and other planning matters are explained in more detail in chapter 6, but a general note may be useful at this stage. Before even an 'outline' planning application can be submitted the applicant will need to know the local authority's views on the following important aspects of a development over which they exercise a measure of control.

Land Use   The first point an inquiry would want to establish is the land use the local authority considers appropriate for the particular site. If this use is in the same 'use class' as the proposed development there is not likely to be any serious objection in this respect but there are a number of other aspects of control to be considered. The Use Classes Order is described in chapter 4 and appendix A.

Residential Density   The viability of a residential development scheme will probably depend on the number of dwellings that can be accommodated on the site. It is important to establish what the local authority has in mind on this point.

Plot Ration and Floor Space Index   If the proposed development is for non-residential purposes the density of land use will be governed by plot ratio or floor space index, depending on which method is used by the particular authority.

Vehicular Access   The local authority has the power to restrict vehicular access to existing roads. Since this restriction may considerably curtail the free use of the site, discussions on this important aspect should take place as soon as possible.

Other Points for Inquiry   The following are other matters that will have to be taken into account and should therefore be the subject of early inquiry

25

Building lines
Daylighting standards
Parking standards
Listed buildings
Tree preservation
Conservation areas.

There may also be additional requirements if the site is in a national park or is one of special scientific interest.

Although a responsibility of central government, the local authority will be able to supply information on the policy being applied locally in respect of

industrial development certificates

Environmental Health

This department of the local authority will be able to provide information on such matters as

(1) policy in respect of street works
(2) location and depth of existing sewers
(3) details of the surface-water drainage system
(4) policy in respect of trade effluent
(5) policy on smoke control and clean air.

The following additional information may also be readily available, and if it can be obtained from this source it may save a lot of time later

(1) soil characteristics in the area generally and on the site in particular
(2) if the site is made-up land, when was it made up and how was the tipping controlled?
(3) information on underground workings and possible subsidence
(4) existing foundations from earlier development now buried
(5) general level of the water table
(6) the possibility of flooding.

This department may also be prepared to furnish a list of contractors in the locality, with a note of the sort of work they normally undertake.

Building Control

The control of building work from the point of view of health and safety is provided for in public health legislation but it is usually treated as an engineer's or surveyor's function. All detailed plans have to be submitted to the local authority to ensure that they conform to the standards laid down in the Building Regulations, and certain stages of the construction are inspected by officers from the authority. (The Building Regulations are dealt with in more detail in chapter 4.) There will be a constant exchange of visits during the process of the development. There is not likely to be very much contact until the detailed planning stage is reached but an early meeting with the officer who will be concerned later is highly desirable.

Other Local Authority Inquiries

Much useful information can be obtained from other departments, but most of this will only be required as the scheme develops. The local authority exercises enormous power in the matter of development in its area. A surveyor concerned with a development, however small, should make a point of visiting the local authority offices as soon as possible after receiving his instructions. This will save time, work and therefore usually client's money.

SITE INVESTIGATION

An early visit to the site is obviously essential in order to obtain a general impression of its setting in the locality and its main physical features, but much of the information required in the early stages of the investigations will be available elsewhere. A much more detailed inspection of the site supported by surveys will be required before a detailed scheme can be produced. The conduct of this inspection and the analysis of the information obtained are discussed in chapter 5.

4 LAND USE AND ITS CONTROL

In the highly complex society in which we live there are all kinds of controls and restraints affecting the use of land, and most of these can be justified for one reason or another. They may be legal and concerned with the ownership of the land. They may be economic. They may be attributable to the physical characteristics of the site. Each of these is dealt with elsewhere in the book. This chapter outlines statutory control over the use and development of land and the buildings which stand on the land. Control over building and development is at present exercised by both central and local government under many different statutes, most of which have their origin in the Public Health Act, 1875. This was one of the first of many important statutes which have empowered local authorities to take steps to improve building standards in the interests of health and safety.

In the past hundred years other legislation has been introduced and there are still many very important aspects of building work subject to control under public health legislation but, since the original 1875 Act, town and country planning has parted from public health and become established in its own right and both housing and highways also have their own legislation. Other forms of control have been introduced from time to time and among these are separate Acts dealing with such matters as factory conditions, fire prevention, police and security, water and other services.

Although these many forms of control are closely related and are often exercised by the same local authority, it is advisable to consider them under the two separate headings of planning control and building control.

PLANNING CONTROL

A book of this kind could not possibly cover all aspects of this subject, but a knowledge of the way in which town and country planning legislation affects and often restricts the use of land is essential when considering any development project.

The 1947 Act introduced town and country planning as we know it, and it was the first Act to include provisions for positive planning. All previous legislation was negative in the sense that proposals were submitted to the local authority and these were either approved or rejected. This Act enabled planning authorities to initiate schemes for development. These powers will be referred to later but it is with the field of control that a developer will be most concerned. There have been changes since the Act came into force in 1948 but many of the original features still remain. This applies particularly to the control of development.

At the time of writing some parts of the United Kingdom are still the subject of planning control under the terms of a development plan which planning authorities were charged with preparing under the 1947 Act. The plan consists of two essential documents: a written statement and a basic map, both of which were submitted to the Minister for approval. These documents, together with other supporting material, explain the planning authority's intentions in the matter of land use and form the basis of planning control in their areas. The Town and Country Planning Act, 1968, provided for the preparation of structure plans supported by local plans and these are now in use by some authorities and as time goes on all county councils will be working from them.

A structure plan is different from a development plan. It is made up of a written statement supported by diagrams and other illustrative documents but it does not include detailed land use maps. It must take into account economic as well as land use needs and will show 'action areas' where comprehensive planning in the form of development, redevelopment or improvement is urgently needed. Ministerial approval is required before a structure plan becomes operative. Under the reorganisation of local government all district councils have become planning authorities in their own right and are responsible for many aspects of planning control in their districts. The control will be exercised by reference to the county council's structure plan. District councils will produce local plans, which must cover action areas and may also cover district plans for larger areas and subject plans for such matters as traffic circulation, the control of mineral working and the reclamation of derelict land. Conservation areas and listed buildings may make up the content of a subject plan.

Each local plan consists of a map with a written statement supported by diagrams, illustrations and other descriptive matter. It may not need ministerial approval.

Apart from a number of specific exceptions provided for in the Acts or covered by statutory instruments, all forms of development require planning consent before work can begin. The word 'development' becomes very important in this context and is defined in the Town and Country Planning Acts as

'the carrying out of building operations, engineering operations, mining operations or any other operations in, on, over or under the land, or the making of any material change of use of any buildings or other land'.

With a definition covering such a wide range of building activities it has been necessary to make an Order to exclude some of these activities, which may fall within the definition, but for which no consent is required. This Order is the Town and Country General Development Order, 1973, as amended by SI 1974/418. The activities include the digging up of the street or other land to gain access to services and also normal building maintenance. The Order also excludes some agricultural land and forestry and most farm buildings.

In dealing with 'material change of use' an Order [the Town and Country Planning (Use Classes) Order, 1972] has been made classifying land uses into different classes; a summary of these is given in appendix A. Any change within one use class does not constitute development.

29

On the other side of the scales certain building activities have been specifically included within the definition so that there should be no doubt that they do constitute development and therefore require planning permission. These are the conversion of a single dwelling into two or more dwellings; the depositing of refuse; and the display of advertisements on external parts of a building not normally used for the purpose.

An application to develop land can take two forms, one of which could be called exploratory, and is known as an outline application. If consent is given to this sort of application more details in respect of siting, design and external appearance of the buildings will be required and these must be approved before work can begin. An outline application is used when there is some doubt whether the type of development proposed will be acceptable to the planning authority. In other cases a full application is made in the first instance and plans are submitted in detail.

There is a statutory procedure to be followed in making a planning application and a time limit is imposed for the planning authority to give a decision. There is the right of appeal to the Secretary of State for the Environment against refusal or against conditions imposed with a consent. This also applies if no decision is given within the time limit when, unless an extension of time has been agreed, the application is deemed to have been refused. This gives the applicant the right of appeal. Conditions can be imposed covering such matters as the number and disposition of buildings, the design, external appearance, materials and type of structure, the location and design of the means of access to the highway and the provision of parking. These and other conditions will also be dealt with later.

An application for industrial development over a certain size (the size is varied from time to time) and outside the development areas may not be considered by the planning authority unless it is accompanied by an Industrial Development Certificate (IDC) issued by the appropriate government department.

There were similar provisions in respect of office development, which required an Office Development Permit (ODP), but this only applied to certain areas and only for a limited period, and the order has not been extended. An application to use land as a caravan site can only be granted if a site licence for the same purpose has also been issued. Advertising is the subject of special control by regulations [Town and Country Planning (Control of Advertisements) Regulations, 1969, as amended by SI 1974/85] and there are regulations for the protection of trees [Town and Country Planning (Tree Preservation Order) Regulations, 1969]. Buildings of architectural or historic interest can be protected by being 'listed' and alterations which will affect their character require a listed building consent in addition to a normal planning consent. Areas or groups of buildings of similar interest can be defined as conservation areas and are given special consideration [Town and Country Planning (Listed Buildings and Buildings in Conservation Areas) Regulation, 1972, as amended by SI 1974/1336].

A number of planning standards are contained in the development plan and if a proposed development does not conform to these standards the application can be refused or conditions imposed requiring these standards to be met. The three most important of these are as follows.

(1) *Residential density* to limit the number of habitable rooms in a given site area.
(2) *Floor space index and plot ration* which are alternative methods used to restrict the floor area of a nonresidential development in relation to the size of the plot.
(3) *Daylighting standards* to ensure that any new development has sufficient light and air and that it does not at the same time reduce that available to existing buildings to below a similar standard.

These standards are referred to in more details in the chapters dealing with specific types of development.

At the time of writing a number of moves are planned to speed up procedures and to improve development control machinery, although the Government does not consider that a review of the development plan system, as recommended by an expenditure committee, is appropriate.

BUILDING CONTROL

The control of building works in the interests of health, safety, welfare and convenience of the public is exercised by local authorities. These are the metropolitan boroughs in the large conurbations, and the Outer London boroughs and district councils in the remainder of the country. A special form of control applies to Inner London, which will be referred to later. A substantial part of building control is by reference to the Building Regulations 1976 as amended, but there are additional powers which are derived from other legislation. Appendix B contains a list of some of these Acts.

The Regulations require the submission of plans and other information concerning building materials, methods of construction and other matters of health and safety, such as resistance to moisture, insulation against fire, heat loss and sound, refuse disposal, drainage, ventilation and structural stability.

The standards adopted by local authorities were for many years covered by local building by-laws. These were largely based on model by-laws published by central government in the 1930s, but over the years there have been changes in emphasis and interpretation and the local attitude towards the many new building materials introduced since that time has resulted in a situation where standards required in one London local authority area could be different from those required in a neighbouring area. The Building Regulations set a national standard and, except for Inner London, apply to the whole of England and Wales. Similar regulations apply to Northern Ireland, and very comprehensive Building Standards Regulations apply to Scotland. A procedure is laid down for the submission of plans and there is also a statutory time limit during which the authority either approves the plans if they meet the requirements of the Regulations or refuse them if they fail to comply. Although in law the authority's reply must be a straight 'pass' or 'fail', in practice building control offices are generally willing to discuss any possible complications before plans are submitted, or during the approval process, so that additional information can be produced or adjustments made. By this means rejections are avoided and considerable time is saved. There is a right of appeal to the magistrates' court, although this right is seldom exercised. There is also a procedure for a joint submission to the Secretary of State for a determination in the case of a particular problem arising out of the application of the Regulations. The Regulations cover all but the very minor aspects of building construction, siting and services affecting health and safety, with only certain nonstructural parts and finishes excluded. The Health and Safety at Work etc Act, 1974, Part III, empowers the Secretary of State for the Environment to make regulations to widen the supervisory and regulatory powers still further. These powers have been utilised to make the (First Amendment) Building Regulations 1978. These came into force on 1 June 1979 and relate to the conservation of fuel and power in all buildings other than dwellings.

The local authority may require specific details of any building work proposed but the following is a summary of the information normally submitted when making an application.

Key plan
Block plan
Detailed drawings showing plans and sections at every floor level, elevations, lowest floor level in relation to site levels and the level of any adjoining street
Position, form and dimensions of all main structural parts, including damp-proof courses, foundations and oversite concrete
Details of class of building and use
Methods of drainage and water supply
Specification of materials
Details of construction accompanied by calculations of loadings and strength

With regard to enforcement the Public Health Act, 1961, states

'It shall be the function of every local authority to enforce Building Regulations in their districts'.

The authority does this by considering plans, issuing decisions on them and carrying out inspection of work in progress. When plans are deposited the authority has up to 5 weeks (or 2 months by agreement) in which to give a decision. If no decision is given the plans are deemed to have been approved, but this does not authorise anyone to carry out work in contravention of the Regulations.

As work proceeds a building inspector visits the site from time to time to ensure that the work is being carried out in accordance with the Building Regulations under which the plans were approved. The local authority must be informed when it is proposed that work should begin and, in addition, at certain stages in the construction they must be notified so that an inspection can be made before work can proceed further. They must also be notified when work is completed and ready for final inspection.

The standards in Inner London are very similar but the powers to control building work arise from a very different source. The principal building Act in force today is the London Building Act, 1930, as amended in 1935 and 1939, and the by-laws in force date in the main from 1964, brought up to date by the London Building (Constructional) By-Laws, 1972. The most significant difference in the administration of building control in Inner London is that it is exercised by district surveyors as statutory officers. They interpret the by-laws and personally take all necessary enforcement action. (Inner London is defined as that part of London which was administered by the London County Council before the Greater London Council was created.)

Note It is anticipated that local authorities will be given the power to make changes for building control services before the end of 1980.

5 SITE APPRAISAL AND ANALYSIS

This chapter deals with that important stage in the development process when an inspection of the site and the adjoining area is necessary for the first steps in the design process to be taken.

Much general information will already have been collected with the original instructions during visits to the local authority and also, in some cases, from discussions with the client's other professional advisers.

The site may well have been visited a number of times and the surveyor-planner may be quite familiar with its general characteristics, but this inspection and an analysis of the information obtained will become essential as the design process proceeds. It will also be invaluable to have all the material properly recorded and in logical order so that it is readily available for reference at any time and, in particular, accessible to other members of a design team or other persons not so familiar with the site.

Any information may be of use and no source should be overlooked, but the really valuable material will be that compiled from a careful analysis of the information gained from a detailed inspection of the site.

SITE INSPECTION

Location

This information will already have been established when instructions were received and during preliminary investigations, but nevertheless it should be recorded in detail as the first item of the report of a site inspection.

Site Boundaries

One of the early tasks in a site inspection must be to identify the boundaries and to note by what means they are defined. If a fence or wall marks the boundary, ownership and responsibility for maintenance should be investigated and if there is any doubt the responsibility should be agreed with the adjoining owners. Although this is not conclusive, it is a useful guide in the case of a boundary fence to note on which side of the fence the posts and rails occur. These are usually on the owner's side of a fence; in the case of a boundary wall the buttresses may indicate the owner's side. This and other typical examples of how a boundary is defined are given in figure 5.1. A wall between the gardens of terraced housing may be a continuation of the partition wall between the houses and as such is a 'party wall', being jointly owned, with owners on each side having certain rights and obligations mainly concerned with support taken from the wall. This is particularly true when it is loadbearing and part of the structure. The centre line of a hedge can be taken

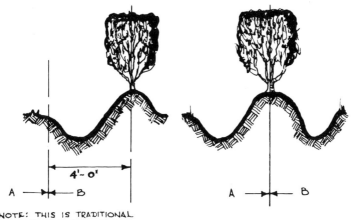

4'- 0"

A ⟶ ⟵ B

A ⟶ ⟵ B

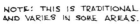

A ⟶ ⟵ B

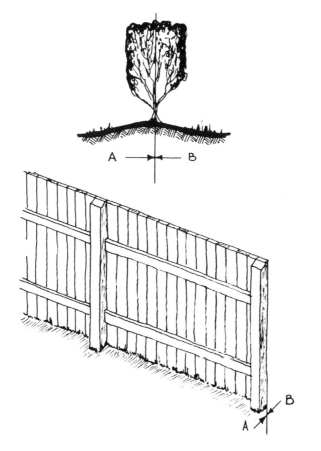

B

A

Figure 5.1   Boundaries

as a boundary if there is no ditch on either side or if there are ditches on both sides. If there is one ditch the boundary is usually 1.2 m from the centre line of the hedge on the ditch side. This rather odd situation is worthy of explanation. Many years ago when an owner dug a trench or ditch to define his boundary this had to be done on his own land. The spoil formed a ridge on the inner side of the ditch and a hedge would be planted on the ridge. Over the years the exact edge of the ditch would become difficult to define so that it has become accepted practice to take it as 1.2 m (4 ft) from the centre of the hedge.

The centre of a river or stream often forms the boundary, except in the case of a navigable river, when the river authority usually owns the river bed. In tidal waters the high-tide line is taken as the boundary. In the case of fast flowing rivers or streams the course may have changed owing to the erosion of one bank and the silting up of the other, which often occurs on acute bends. Boundaries with public thoroughfares or other land to which the public has access should be especially noted. Although the land is not strictly part of the site it may be useful to note the land use immediately beyond the boundary.

Adjacent Roads, Streets and Other Rights of Way

A new vehicular access to the site will invariably be subject to stringent conditions imposed by the highway authority and therefore any existing access should not be lost by default or by not knowing of its existence.

The orientation of individual buildings within the site will often be influenced by existing roads, paths and tracks. All these should be carefully plotted on a plan and notes made of their construction, classification and usage. These notes should be particularly detailed where there is already established access to the site. Levels of the site are dealt with later, but a note in this part of a report would be useful if the level of the site differs widely from that of the surrounding land.

Railway Tracks

These can complicate a development by forming a barrier between two parts of the same development scheme. If the line is busy it may constitute a noise nuisance. In this connection it is important to note whether the track is on an embankment or in a cutting.

Rivers and Streams

Watercourses can either be exploited and made an attractive feature of the development or they should be diverted or controlled so they do not interfere with the maximum use of the site. Their exact course should be plotted and a note taken of the volume of water and the rate of flow. It is important to note the time of year when the inspection is made because there may be substantial seasonal variations which have to be taken into account.

Site Levels

An instrument survey will be essential at some stage in the planning of a scheme but, in most cases, for the purposes of this early inspection, sufficient information can be obtained by reference to local maps showing contours, trig points and spot heights, coupled with the visual inspection.

Levels will have some influence on the siting of roads because steep gradients are to be avoided. They will also need to be taken into account when the sewage system is being planned, and they are very important in land drainage.

The siting and orientation of individual buildings will be influenced by levels but this particular aspect of site planning will be dealt with later in this chapter under 'points of vista'.

Water Table

The level of saturation below the surface may affect plans for earth moulding and landscaping. A high water table is taken into account when designing and specifying materials for foundations.

Levels which are likely to cause problems later may be lowered by improved systems of land drainage. The usual method of measuring the water table is to dig holes at selected points on the site and leave them overnight so that water can percolate into them to find its own level. As with surface water this is likely to have seasonal variations.

Other Physical Features

Mounds and depressions may be man-made or they may be part of the natural formation of the land. Natural mounds may be incorporated into a scheme to show off prominent buildings. Depressions can sometimes be used to partially hide unattractive buildings. Man-made mounds may be barrows or ancient burial grounds, in which case they are protected for their archaeological interest. Depressions in mining areas may indicate subsidence. Ponds and lakes can sometimes be incorporated into the design to give character or they may be filled if they do not form part of a watercourse. Flood-plain gravel workings fill naturally with water and the banks are often landscaped, with the water left to provide for fishing and sailing and other water sports and pastimes. Dry quarries and pits can sometimes be filled but strict rules are usually laid down by the local authority to control the material to be tipped, the rate of tipping and the after-use of the site. The exposed face of a stone quarry may provide a dramatic feature around which to plan a development.

When considering physical features generally a good rule is to exploit them if they can be used to add character to a development scheme, but if not they should be neutralised or hidden.

Pits should always be plotted on a plan and rough calculations made of their capacity so that when refilling is necessary steps can be taken as soon as possible to begin operations.

Existing Trees and Shrubs

Whenever possible mature trees and shrubs should be retained and incorporated into the landscaping of the new development. They should be plotted on a plan usually to a scale of 1/500. This can then be reproduced on a sheet of transparent material so that it can be used as an overlay to check positions when the exact location of roads, buildings and other development is being decided.

Trees should be classified into one of the following four groups and the appropriate letter placed in the centre of the tree as plotted on the plan

A Valuable   to be retained whenever possible
B Desirable   to be retained unless their retention makes the project unworkable
C Inessential   may be retained or not, as planning and other requirements dictate
D Undesirable   should be removed owing to disease, damage overcrowding or similar reasons.

The following additional information should be noted for future use in landscape plans

> Species
> Approximate height
> Approximate spread
> Level of lowest branches
> Diameter of bole (if above 150 mm)
> Shape and growth characteristics
> Condition.

More information on this important subject is given in chapter 14.

Soil and Subsoil

It is most unlikely that all the detailed information required about the soil and subsoil will be available from a first inspection of the site. Trial borings, chemical analysis and other technical procedures will have to be followed and most of this will be done later when loadbearing requirements of different parts of the site have been calculated.

Certain basic information will be required when the plan for the scheme is first being formulated and the most straightforward way of providing this information is to classify the site or parts of the site according to its likely condition.

Natural State When only the topsoil has been disturbed by being used for agriculture the make-up of the subsoil is likely to be known locally and information will be available from the results of archaeological surveys.

Developed Land   Where there are existing buildings, whether they are still in use or are derelict, it is generally possible to estimate the disturbance of the subsoil. In any case the demolition of buildings can be supervised with an eye to the stability of the subsoil for future development. In cases where buildings have been demolished and the surface has been levelled the disturbance will be more difficult to trace and reference to earlier plans and a detailed site investigation will be essential to discover where the original foundations were and to see if they still remain and were just covered over when the site was levelled.

Made-up Land   Land which has been made up after mineral operations should be checked to establish what kind of filling material was used. If the filling was started after 1949 the planning authority will have a note of the conditions imposed with the granting of planning consent to tip. If tipping took place before planning control was introduced this may create a problem.
    So far as it affects future land use tipping can be subdivided into three kinds as follows.

(1) *Builders' rubble*: this is the best type that is usually available. If the tipping is strictly controlled and the land is allowed to settle and consolidate it can be safely used for most forms of development.
(2) *Household refuse*: this is less satisfactory and often has a long-term effect on the possible uses of the land afterwards. Planning authorities generally impose restrictions on its use after tipping is completed.
(3) *Toxic and other chemical waste, industrial waste*: this may so contaminate the land that it is of little use for development for many years after tipping has ceased. When development is permitted, elaborate precautions have to be taken to ensure that the chemicals do not adversely affect the materials used in the construction of the foundations.

Land with Underground Mine Workings   Land in this category is always liable to subsidence, particularly if the workings are old and near the surface. It is important to appreciate that, although modern workings are carefully plotted by the mining company, those worked in the distant past are less likely to be so accurately recorded and they are the ones that are often to be found nearest to the surface. Their shallow location and the fact that pit props and other supports may have deteriorated over the years makes subsidence a constant possibility.

Existing Buildings and Other Land Uses

At this stage a note should be taken of the uses of all the land and buildings still in occupation so that plans can be made to terminate existing tenancies and if necessary provide alternative accommodation. This will ensure that there is no delay when the redevelopment is ready to begin.
    Buildings that are not to be demolished because they are protected as listed buildings or for some other reason will have to be subject to a structural survey. This may not be convenient or even practicable at such an early stage.

A careful record should be made of their presence on the site and their exact location should be plotted. If there are a number of protected buildings an overlay similar to that used to show existing trees can be prepared but if there are just one or two they can usually be added to the overlay for trees.

Points of Vista

Much will need to be done in this field as the scheme develops. This report should include a description of the surrounding area from a number of predetermined points on the site (usually high spots) to help when considering the layout of the individual buildings in relation to other buildings. This is done with the object of leaving attractive views free from obstruction and at the same time hiding the unattractive ones.

A similar process can then be carried out the opposite way round by going beyond the boundary to points where the site can be seen. This can be taken into account in planning the development to look its best from where it will be most seen.

Adjacent Land

The land immediately surrounding the site will have a considerable influence in the planning for the development of the site whether the proposed scheme is new or part of the process of urban renewal. A survey of this land should be carried out in two phases. It will be necessary first to produce a factual report on land uses as they exist with general comment on and condition of any buildings.

The second phase should be the investigation into any development proposals for the area. If they are to take place within a reasonable period these may give an indication of how the locality generally is likely to change. This may have more influence on the proposals for the site than the surroundings as they exist at the time of the survey.

TECHNICAL INFORMATION FROM OTHER SOURCES

Much useful information can be obtained from the local authority, whose functions result in its having information on planning, public health, housing, education and many other aspects of life in the locality. Such information must have a bearing on any proposed development within its area.

TYPE OF DEVELOPMENT

Even at a very early stage in any proposed scheme the type of development being considered is likely to be known and the site inspection must be influenced by this knowledge.

This chapter has covered investigations and a site inspection of a general character but more details may be required according to the type of development proposed. References are made to more detailed surveys specifically directed towards the particular types of development in the chapters devoted to them. Chapters 8, 9, 10, 11 and 12 each deal with a different type of development and contain references to additional information where this is needed to meet a particular requirement.

6 THE ECONOMIC APPROACH TO DEVELOPMENT

Most of the students reading this book will already have studied the technique of valuations in other parts of their course. This chapter is not intended as a substitute for those studies but rather as a broad look at the financial aspects of development, which must include some reference to valuations.

Some development schemes are not judged solely on their economic viability or on the financial return they show for the developer. Many local authority schemes, for example, can only be justified on social grounds. A private developer, on the other hand, must show a profit, or at least a potential for profit, at some not too distant future date. In the case of a development of, say, a town centre undertaken jointly by the local authority and a private developer, it may incur both social and economic costs and provide both social and economic benefits. This chapter examines generally suitable techniques for assessing values in the form of financial return, economic gain and social benefit, but it does not take account of any speculative interest in land for development.

The author has always been of the opinion that the role of the developer should be considered separately and not be combined with that of a speculator. It is not suggested that there is anything wrong with speculation but only that it should be thought of as a different commercial activity.

A developer who has bought land either with planning consent or even with a planning potential has acted wisely. To create a land bank is the only way to ensure continuity of work, and this is not so much speculation as good management. In ordinary circumstances the starting point when considering the financial aspects of a development should be the site, at a figure based on its value with the appropriate planning consent. This would have become normal practice had the Community Land Act 1976 stayed on the statute book long enough for it to take effect. It was operated for a number of years but at a very low key and it is now virtually dead. It was designed to enable the State to benefit from most of the increase in land values directly attributable to the granting of planning consent. In a hypothetical example of how the scheme should work published by the Ministry, the tax on a £200 000 sale of land was £127 000 after deducting the value of the land without consent and taking into account other allowances. This is approximately 64 per cent of the selling price. This levy or taxation has become known as betterment and there have already been two major attempts by the State to collect all or a large proportion of what is often substantial benefit directly resulting from a planning consent.

The first was contained in the Town and Country Planning Act, 1947, which provided that from 1 July 1948 there should be a development charge when planning permission was given for any development other than 'existing use' development. The amount of the charge was based on the difference between

(1) the value of the land with the benefit of planning permission and
(2) its value with permission refused.

There was a compensation provision in respect of any development value which existed prior to the above date and this was to be paid out of a global fund of £300 million allocated for the purpose. The Central Land Board was established to levy the charge and pay the compensation. This action by the government had the effect of almost drying up the supply of land for development and this part of the 1947 Act was partially repealed in 1954 and fully repealed by the Town and Country Planning Act, 1959.

The second attempt was the introduction of the Land Commission by an Act of that name in 1967. The Commission had two completely separate functions, one being to stimulate the development of land and to use compulsory purchase powers to do so if necessary. The second, and the one for which it was given the most publicity, was to levy a charge for betterment. To avoid the imperfection of the 1947 attempts to collect betterment, this Act provided for types of betterment to be grouped into six classes (cases A to F). The Act came into effect on 6 April 1967 and continued until 22 July 1970, when the Government announced that legislation would be introduced to abolish betterment levy on any future transactions.

The only part of this controversial practice which has remained is the right of the acquiring authority to set off against the value for compensation on a property being acquired, the amount of any increase in the value of any adjoining land in the same ownership. This increase must be directly attributable to the authority's scheme. There are also small amounts of compensation which are relics of the compensation provisions of the 1947 Act. These are still attached to the land and in certain circumstances their value can be deducted from compensation should the particular land be the subject of compulsory purchase. This is known by the grand-sounding name of 'unexpanded balance of established development value'.

It is likely that there will be further attempts to collect betterment by extending the provisions of capital gains legislation. Whether this happens or not it should always be assumed during the financial appraisal of any scheme that the value of the site is taken as market value with the appropriate planning consents. Any speculative rewards which might have been gained by obtaining planning consent should be ignored when considering the viability of a scheme.

VALUATION OF A SITE FOR DEVELOPMENT

Market value can sometimes be more and sometimes less than a developer is prepared or is able to pay for the site in order to carry out a proposed development. Under these circumstances he will need to know the maximum amount that he can afford to pay for the site in order for the scheme to be viable. This is done by a valuation technique known as residual or development valuation, and, as its name implies, it is a method which works as follows.

First take the estimated capital value of the completed project. From this figure are deducted

(1) the cost of demolition and site preparation
(2) the estimated cost of site works
(3) the estimated construction cost
(4) interest charges on the capital involved (other than land)
(5) fees to be incurred in respect of professional advice, negotiations and supervision of work concerned with the building; fees also to include those for the sale or lease of the completed development but not for the purchase of the land
(6) legal charges connected with the disposal
(7) developer's profit.

The figure remaining is the maximum the developers can afford to spend on the site. This is not just the purchase price but includes the fees connected with the purchase and the interest on the land during the time it is kept fallow and therefore unproductive. In cases where parts of the site are still occupied, compensation and other costs will arise and these will also have to be deducted from this item.

It is important not to lose sight of the interest element in this part of the calculation. It becomes operative immediately when the land is purchased and may run until the new development is completed and sold or let. This applies particularly to large office blocks and the like. It is slightly different in a residential development of houses to be sold freehold since the cost of the land begins to return when the first house is sold. Interest in this case can be best calculated by reference to a 'flow chart', referred to later in this chapter. This valuation method should only be used in situations such as that described. The Lands Tribunal has cast doubts on the method's suitability for other valuation purposes but, used for the purpose for which it is designed, it is quite satisfactory and is the most realistic method that can be adopted.

DEVELOPER'S BUDGET

In situations where the value of the site has already been agreed, a similar exercise but worked in a different order will provide answers to most of the questions on costing that may arise when testing the viability of a proposed scheme. It will indicate where savings have to be made if costs are too high. This type of exercise is generally know as the developer's budget.

DISCOUNTED CASH FLOW

There are constant references to cash flow problems when the financial stability of a firm is being assessed. This usually seems to suggest that the firm has capital assets but there is a shortage of liquid funds for immediate use. A situation such as this can easily happen in the field of property development, and at worst it could probably result in the financial failure of the whole project. At best it suggests that additional working capital has to be found at short notice but this is likely to be at a much higher rate of interest than was budgeted for. It will usually have many strings attached and often contain a financial commitment which results in most of the anticipated profit being creamed off.

Discounting is a perfectly straightforward arithmetical procedure, based on the simple premise that money is only worth its full face value if it is immediately available, as the following simple examples illustrate.

(1) An undertaking to pay certain sums of money in two years' time is not worth its face value today. Its maximum value is its face value *less* what it would earn in the way of interest in those two years.
(2) Conversely, in order to meet a contractual undertaking to make a payment in two years' time there would need to be available at the present time, not the full amount of the payment, but a sum which would earn sufficient interest to accumulate into the full amount in the two years before payment is due. The calculations are routine and can be looked up in valuation tables, but the important part of the exercise is the rate of interest to be adopted and this will depend on the many factors associated with a particular scheme.

Discounted cash flow techniques have two main functions. Firstly, they provide a realistic assessment of the profit margin when payments are made at certain stages in a scheme and particularly when there are receipts from the sale of parts of the development before the scheme is completed. A good example of this is in a residential estate where houses are sold off as they are completed and ready for occupation.

The second function is one that reveals the amount of money that will be needed at any stage in the development. It enables the time scale of the scheme to be related to funds available and to the cost of those funds in the way of interest. As a general rule the availability of funds dictates the programming of the scheme.

COST BENEFIT ANALYSIS

Development projects undertaken by local authorities or other public bodies are often difficult, if not impossible, to evaluate by traditional valuation methods. The principal aim is not profit and the normal price mechanism does not usually apply. Cost benefit analysis provides a technique of evaluation and appraisal which takes into account not only the economic cost but also the social cost and benefits. The technique has been used at a national level on many occasions, including the appraisal of the M1 motorway, the Victoria line in London and London's third airport proposals. It has also been used to decide which of a number of alternative proposals for road improvements should be used.

A technique for comparing alternative schemes, developed by Nathaniel Lichfield,[1] overcomes some of the inherent difficulties in trying to equate economic or financial benefits or costs to purely social ones. It takes the form of a planning balance sheet in which costs and benefits are recorded and calculated in different denominations. Known money costs and benefits are recorded as such but, where a specific financial value cannot be attributed to a cost or benefit, algebraic symbols are used and this applies to most of the social costs and benefits.

This technique has its uses but it does not give all the answers. It can be used to advantage when evaluating a number of alternative methods, each designed to meet a specific objective, but it can be dangerous when attempts are made to place monetary values on social activities in order to produce a tidy arithmetical answer. It should be treated as a tool to assist the decision-makers but not as an answer in itself. All the known costs and benefits of the scheme to be compared should be marshalled for analysis. They should then be recorded on the balance sheet. Where like can be compared with like a conclusion can be drawn, but not otherwise. The result should be such that all the advantages and disadvantages of each of the alternative schemes can be readily seen so that a decision can be taken on the information provided.

HOUSING COST YARDSTICK

This plays an important part in the economics of the local authority scheme of residential development. In order to obtain the approval of central government to a housing project, and thus be eligible for subsidies and loan sanction, a local authority has to plan within the cost yardstick laid down. This is a limitation on building costs measured on the basis of housing one person. The cost limit varies for different residential densities and for different mixes. There are also slight regional variations. A procedure exists for local authorities to make special application to exceed the yardstick where local circumstances make it impossible to keep to it in a particular scheme.

PROPERTY FINANCE

There are six basic ingredients for a financially successful development.

(1) The site should be well located and suitable for the proposed development.
(2) The site should have been purchased at the right price.
(3) The development should be suitably designed to satisfy known demand.
(4) Sufficient expertise should be available to carry out the development.
(5) The scheme should be practicable within a planned time scale.
(6) It should be correctly financed: this is probably the most important factor.

Three main sources of finance are usually available to a developer and their use is dependent on the time factor.

Short-term Finance This is generally available over 2 or 3 years and is normally obtained from one of the joint stock banks at rates of interest of from 2 to 4 per cent above base rate. (The Bank of England exercises control over the lending policy of joint stock banks and, in certain economic conditions, issues a directive restricting lending in any particular field.) The site will be taken as a security and the bank may need other securities in the form of charges on other properties and guarantees. This type of bank does not ask for an equity share in the scheme. Merchant banks are often a ready source of short-term finance and they tend to be more adventurous than the more conservative joint stock banks. However, they do charge higher rates of interest and expect an equity share. Some of the larger property companies are often willing to lend on short term when the market is stable, and they will sometimes take an equity share.

Medium-term Loans These are for a term of up to 10 years. Medium-term loans can normally be obtained from the same sources as short-term loans but interest rates are usually high and fall midway between those for short and long-term loans.

Long-term Loans These can take a number of forms but funds usually come from the financial institutions.

(1) *Mortgages* These may be for a period of 20 to 25 years but the interest rate may be quite high. As a form of investment it is not particularly popular with the institutions since it does not provide for a share in the equity. The interest until the first rent review may be higher than the return on capital but if a developer can survive during this period he benefits by retaining the freehold intact and enjoys the full benefit of any capital growth. The loan is usually restricted to an amount not more than two-thirds of the total value of the property, on which it is secured.

(2) *Equity participation* This method is not very much used because of its complexity. There is a minimum rate of interest plus a share in the increase on rentals at each rent review. The advantage to the developer is that it reduces the possibility of a shortfall between the return of capital and the interest charges before the first review but on the other hand he does not enjoy the full benefit on the increase in rents at the review.

(3) *Sale and leaseback* It is sometimes said that this was first introduced in the late 1950s but although it became much more popular around that time it was used in the 1930s. The method involves an institution purchasing the property and then advancing the total development cost to the developer, who uses the funds to carry out the scheme. When the scheme is completed the developer is granted a long lease on the whole property and he in turn leases individual properties to the occupiers. When this method of financing a development was first used there was generally a provision for rent on the head lease to be increased by fixed amounts at certain stages during the term of the lease, but it is now the usual practice to provide for occupational reviews, which means that every time the rent is increased the head rent is increased proportionately.

(4) *Mortgage-linked leaseback* This is a variation of the normal sale and leaseback procedure, the difference being that it combines the granting of a mortgage with leaseback.

47

THE DEVELOPER'S BRIEF

The building and development industry is very large and complex in its structure. It is made up of firms of all sizes and the grouping together of specialisations and skills within one organisation varies from one firm to another. Some firms are big enough to carry out a major scheme using only their own resources but in most cases a number of firms work together, each contributing a particular expertise.

A development scheme can be broken down into five fairly well-defined phases

> Site acquisition
> Provision of finance
> Planning and processing the scheme
> Construction work
> Disposal.

Each of these phases calls for expert knowledge and the object of a brief should be to link them together to produce a successful project.

There are many different relationships between those engaged in a scheme and communications between them must be soundly based to ensure a free flow of essential information. The variety of these relationships and the different financial interests involved show the importance of a master plan to outline the scheme in sufficient detail to form the basis on which detailed instructions can be given and specifications prepared. It should remove doubts and uncertainties and should be the authority to which reference is made if disputes arise. A fully comprehensive developer's brief could be the basic document of this master plan.

In the building and development industry the word 'brief' is used in different ways and briefs are used for a number of purposes. A client's instructions to his architect are sometimes called the architect's brief. A contractor is sometimes given a brief instead of a specification. This usually occurs where parts of a scheme are not planned in detail but where the contractor is invited to come forward with ideas and suggestions. The size and content of the brief will depend not only on the size and complexity of the scheme but also on the stage at which the brief is prepared, the phase of the development it covers and the extent to which design and financial control is to be exercised by the person writing the brief.

The Department of the Environment has published advice on the preparation of a developer's brief for private residential development.[2] It was issued as a guide to local authorities needing to initiate private residential development on sites acquired under the Community Land Act, 1976. The Department wishes to ensure that developers are given scope to innovate and to apply their skills and knowledge of the market, but, at the same time, that they will be concerned with the quality of the environment, the costs of development, the preferences of home buyers and the prices they will be able to pay. A brief is seen by the Department as a way of speeding up a development and avoiding unnecessary negotiations or delays during the design and approval stages. The Advice Note includes the following very useful check list.

(1) Would the preparation of a brief be an advantage for the disposal and development of the site?

(2) Can it be limited to a planning permission and the relevant details of the authority's development control policies?

(3) If more is needed, what are the essential requirements, for example, for layout and appearance, and what can be left to the developer?

(4) Will there be any need to give publicity to the content of the brief?

(5) Are any major requirements by other authorities indicated?

(6) Are any conditions and restrictions on either the site or the developer clearly specified?

(7) Does the brief include all the basic information the developer needs?

(8) Have the implications for costs, timing and marketability been taken into account?

(9) Is the brief concise?

(10) Has the authority taken steps to ensure that the views of potential developers are sought on its approach to the use of briefs, and that suitable arrangements exist for monitoring?

The make-up of a brief for a council housing estate is given as appendix C.

REFERENCES

1. Nathaniel Lichfield, *Economics of Planned Development* (Estates Gazette, London, 1956)
2. Department of the Environment, *Development Advice Note No. 1* (HMSO, 1976)

7 ROADS, STREETS AND FOOTPATHS

A developer will be concerned with roads and streets and possibly footpaths. He may have to construct them as part of a development scheme. He may be responsible for all or part of the cost of construction. He may be required to contribute towards the cost of making up an unmade road. He may wish to close or divert an existing road, street or footpath and he will certainly be concerned with vehicular access to roads generally. He may enjoy the right to use roads owned by other people or he may be compelled to allow other people the right to use roads owned by him. Many words are used by the layman to describe a road or street, but a developer and whoever he consults should know exactly what both words mean. A number of definitions may be helpful.

(1) A road to which the public at large has by law the right of access is called a highway.
(2) At Common Law highways are ways over which every member of the public can pass and repass free from restrictions unless the restrictions are imposed on the public as a whole.
(3) The Highways Act, 1835, extends the definition of highway to include bridle-paths, bridges, causeways, footways and the like.
(4) 'Street' has a more recent statutory definition but it is incomplete in that it extends the types of way which fall within the definition without first defining 'street'. The most significant part of this definition is that a street need not be a thoroughfare (an unobstructed way).
(5) In the sense that the word is used in the Street Works Code it can be taken as meaning a road, etc., which is or is likely to be flanked by buildings. (An outline of the provisions contained in the Street Works Code is given in appendix D.)
(6) The Buchanan Report [1] also contains a useful definition of a street: a form of layout consisting of a carriageway for vehicles, flanking pavements for pedestrians and with frontage development with direct access to premises for pedestrians and occasionally for vehicles.

Because of the great age of many UK roads, ownership can be a complicated matter and a developer should not fail to consult his solicitor if he is in any doubt; however, for the purpose of this book they can be taken as falling into the following two categories.

Roads and Streets Maintainable at Public Expense

These form the majority of all roads and streets but they are not usually owned by the public. The public has the right of way over the surface, and, in order to be able to maintain the road, a sufficient depth of soil is vested in the public.

The land over which the road or street runs remains in private ownership, usually that of the person who owns the land fronting on to it.

The rights to lay sewers and other services under the road surface and the right to plant trees on the grass verge are contained in other legislation.

Although fewer by comparison, a growing number of roads, including most motorways, are wholly owned by the public, the land having been acquired for the purpose of creating a new road or substantially improving an existing one.

Private Roads and Streets

These are in private ownership and are not maintainable by the public. The owners are generally the frontagers. They may be residential roads which for a number of reasons have never been adopted or they may be service roads forming part of a new development.

The maintenance of roads for which the public is responsible is carried out by the highway authority and is different according to the type and classification of the road concerned. Central government is the highway authority for trunk roads. Classified roads are maintained by county councils, together with all unclassified roads, unless the appropriate district council has claimed the right to maintain unclassified roads in their area.

Although private roads may be quieter and more secluded than public roads there is often difficulty in organising repairs and getting all the frontagers to contribute towards the cost. This difficulty is sometimes overcome where owners wish to keep their road private by forming a limited company for the purpose of maintaining the roads. This action does not guarantee that the road will remain unadopted by the local authority, but it could remain so if the road is properly maintained.

Owing to difficulties over maintenance some residents prefer to have their roads adopted and, provided the condition of the road is up to the standard required by the local authority, this can often be done at no expense to the resident. In most cases the condition of the road is substandard and the work is carried out by the local authority and the cost apportioned. A similar procedure is followed if the local authority decides that the road should be adopted and applies the Street Works Code.

There is a special procedure to be followed when development takes place on land fronting on to an unadopted road and where there are indications that it is likely to be adopted within a reasonable time. (This procedure is known as the Advance Payments Code and is outlined in appendix E.) This takes the form of a payment or a formal undertaking to the local authority to cover an estimated proportion of the cost of the work when the road is adopted.

STOPPING OR DIVERTING AN EXISTING ROAD

Under highway legislation it is possible to stop or divert a highway by an application made to the magistrates' court by the highway authority. If a developer wishes to apply he can only request the highway authority to do so on his behalf. There is no right of appeal if the highway authority refuses to do so. The magistrates may authorise the closure if it can be shown that the road is unnecessary, or its diversion if this can be done at no disadvantage to the public.

In a case where slum clearance is being carried out by a local authority, roads can be stopped or diverted under housing legislation. In an urban development by a government department or by a local authority, town planning legislation can be used.

NEW ROADS

The new roads with which a developer is likely to be concerned are those whose main function is to provide vehicular access to the properties that make up a new development. In the case of residential and industrial developments these will generally be the estate roads. In town centre development the developer will be concerned with service roads to the rear of shops and roads serving car parks. The all-purpose roads in the town centre will, almost invariably, be the direct responsibility of the local authority.

The design and construction of roads forming part of a new development are usually the responsibility of the developer and are carried out in two parts

(1) The layout of the road pattern to meet the requirements of access and to allow for landscaping
(2) The structural drawings and specification for the different types of road within the estate.

These are generally carried out by a consultant engineer after he has met the officers of the local authority who will ultimately be responsible for taking over the roads when the development is completed.

The local authority has the power to stipulate level, width and construction of new streets and can require plans and sections to be submitted before work is started. It is normal practice for the developer to construct the roads required for new residential or industrial development and to hand them over to the local authority when the development is completed. These new estate roads are maintained by the developer until a fixed period of time after vesting day (the day the roads are adopted by the local authority). At the expiry of this additional period of responsibility and if the roads are still in good order they become maintainable at public expense.

FOOTPATHS

Footpaths over development land can often restrict the development of that land. Local authorities are required under the National Parks and Access to the Countryside Act, 1949, to keep and make available to the public a map showing all the footpaths in their area. Each footpath is registered and is given a number.

It is most unlikely that an application to close a footpath would succeed since there are a number of very active organisations determined to keep them open to the public. In rural areas the parish council has to be consulted. A footpath can generally be diverted if a case can be made for its diversion and if, at the same time, it can be shown that the proposed alternative route is no longer, or any less pleasant as a walk, than the original one.

CLASSIFICATION OF ROADS

Roads are classified by central government for administrative purposes and for purposes of grant aid to highway authorities in the following manner

Motorways	national through routes
Class I	arteries of national traffic value
Class II	arteries of regional importance
Class III	roads of local importance
Unclassified (district)	roads of local value only.

All but the unclassified rank for grant aid on a sliding scale related to their importance. Since it is the unclassified road that is likely to be the concern of the developer, he can look for no financial assistance in the form of a government grant.

Alker Tripp, Deputy Commissioner of the Metropolitan Police in the early 1940s, was probably the first to draw attention to the changes taking place in the road usage owing to the large increase in the number of motorcars. He pointed out that the all-purpose urban road was becoming out-dated. We could no longer expect roads to provide individual vehicular access to buildings fronting on to them, and, at the same time serve as through traffic routes from one part of a town to another. These views were amplified by others, including Colin Buchanan,[1] and together they led to the publication of a manual by the Ministry of Transport[2] in 1966 in which roads were classified in relation to their function as

(1) primary distributors, forming the primary network for a town on to which all-long distance traffic moving within the town should be channelled
(2) district distributors, to distribute traffic within the districts of the town
(3) local distributors, to distribute traffic within the environmental areas
(4) access roads, to give direct access to buildings and land.

A developer will be mainly concerned with access roads but in a large development he may also have to provide local distributors. A general name given to what are called access roads in the above classification is 'minor roads'—roads which actually serve the houses and other buildings. They should have no other function except to carry local traffic to the nearest spine road (similar to local distributor in the above classification).

SITING OF ROADS

The provision of vehicular access to most forms of development is vital for its full social enjoyment and economic use. For this reason roads are a very important aspect of any modern development. In cases where a complete road pattern is already established and in full use it is often difficult to make substantial changes, but in a development where new roads have to be provided they should be sited to serve the development. In an ideal situation the most suitable position for the buildings should first be selected and then vehicular access provided by bringing a road to the buildings. This ideal is seldom attainable in modern developments for a number of reasons, the main ones being

(1) construction costs
(2) the need to utilise both frontages whenever possible
(3) the disproportionate amount of land which would be used for roads, and
(4) the high residential densities currently acceptable.

When considering the siting of a road it should be borne in mind that it consists of the carriageway, the footways or pavements and the grass verges, if any. The siting of roads in relation to the different types of development will be dealt with in more detail in later chapters devoted to different types of development.

ROAD DESIGN AND CONSTRUCTION

The construction details generally and the dimensions in particular will vary considerably according to the type and intensity of the traffic using the road. There are, however, two fundamental factors which have to be taken into account when planning a new road for whatever purpose.

(1) The first is concerned with the geological formation, the types of soil and the level of the water table. This information is obtained by boring and other survey techniques and the results are usually recorded on a soil profile accompanied by a key to soil types, a table of test results and a sectional drawing showing the level of saturation. This information is essential when the loadbearing capacity of the road is being considered.
(2) The second factor to be taken into account is the topography of the site. This is important when considering gradients in relation to the type of traffic likely to use the road.

These two factors are concerned with the physical characteristics of a site and the approach should be to make the best use of those helpful to the scheme and to overcome or minimise the effects of those which are likely to be unhelpful.

The remaining factors may be more appropriately called design considerations. They sometimes make use of the physical characteristics but are generally introduced into the design.

(3) The safety of those using the road must be a prime consideration in road design at all stages and for all types of road. Because the scope of this book is limited to access roads, with the possible inclusion of local distributors, the safety factors applicable to these minor roads only will be considered.

(a) The practicability of the segregation of vehicles and pedestrians should always be considered.

(b) The number of road junctions and intersections should be related to the functions of the road.

(c) The radius of the kerb at the junction should permit traffic to negotiate it safely and should be not less than 6 m.

(d) There should be adequate sight lines at road junctions. The usual minimum standard is at 1 m above ground level for 60 m along the kerb of the carriageway of the major road from a point 4.5 m down the centre of the minor road.

(e) The number of individual vehicular accesses allowed should take into account the number and types of vehicle using the road.

(f) Business premises fronting on to a busy road should have loading and unloading facilities off the road.

(g) Culs-de-sac should have an adequate turning area at the end. A circular head should be not less than 16 m in diameter and a T-head not less than 10.5 m. The radii of all the curves into the head should be 6 m.

(h) Premises attracting visitors should have ample parking space off the road.

(i) A building line should be defined in respect of all roads with frontages which are likely to be developed. Between 6 m and 8 m is generally acceptable.

(j) Footways should be of adequate width and unobstructed by street furniture.

(k) Street lighting should be adequate and uniform and lamp standards should be not closer than 0.45 m from the carriageway.

(l) The surface of the carriageway should be nonskid and that of the footway nonslip.

(m) Steep longitudinal gradients should be avoided, particularly at road junctions. These should be not greater than 1/50 within 7 m of the junction.

(n) The camber and crossfall should be correct for the type and surface of the carriageway, usually between 1/48 and 1/36.

A number of factors will arise in later chapters dealing with detailed layouts of sites for different types of development.

(4) The appearance of the road is of importance to those occupying premises overlooking the road and also to other people using the road. It will be largely a matter of siting the road in relation to any physical features and trees that already exist and the preparation of a landscaping and planting scheme to fill in the gaps

(5) The amenity value of the road is clearly linked with appearance but may also embrace convenience and matters concerned with the frontagers and other users of the road.

(6) The prime function of minor roads should be to provide access to buildings fronting on to them. The interest of the frontagers should be treated as one of major importance.

(7) The dimensions of the carriageway and the footway should be related to

the type and amount of traffic using the road and the provision of grass verges should be considered in the same context.

Design Standards

There are minimum recommended standards for most types of road (see figure 7.1), which must be observed when new roads are being planned. Essex introduced some modifications to residential roads to apply locally but it was not until mid-1977 that some flexibility was introduced into national standards, when a new bulletin was published[3] which should result in a reduction in the proportion of land taken for roads and footpaths, thus making higher residential densities possible without undue loss of privacy.

The emphasis is on the creation of a roads and footways network appropriate to the traffic it will have to bear. The bulletin explains how road widths can be reduced, curves tightened and splays at road junctions made narrower. In order to attain this reduction, nonaccess traffic must be routed away from dwellings and there should be no access from distributors. Access roads should be designed as a loop or cul-de-sac off the distributor roads and flow can be reduced by keeping these as short as possible.

Speeds can be reduced in a number of ways, including the introduction of mini roundabouts, tightening bends, avoiding long straight stretches, having changes in the road surface and buildings closely spaced. Street parking must be reduced and wherever possible parking by residents should be within the curtilage of the dwelling. With regard to the width of the roads no hard-and-fast rules are laid down but they should be related to the characteristics of the vehicles likely to use the road. Recommended access road widths vary between 5.5 m, which will allow all vehicles to pass one another, and 3 m, for single track roads with passing places. There is also provision for pinch points down to 2.75 m in width to reduce speeds and to allow for pedestrian crossings. Curves at junctions should be determined by the classification and the flow of traffic expected and may be as tight as 4 m radius for roads carrying light traffic, even if the occasional long vehicle has to use the width of the whole carriageway at the curve. The minimum distance between road junctions on access roads has been reduced to between 30 m and 40 m from 90 m (about 300 ft) but the distance on distributors remains the same.

Components in the Construction of a Road

The Carriageway   This consists of the following structural elements

(1) *Wearing surface*   this is formed in a number of different ways and can consist of water-bound macadam which is an old-fashioned but still satisfactory way of forming a road surface. It is now usually treated with a dressing of tar and chippings. Tarmacadam differs from water-bound macadam in that the stone is coated with tar before it is laid and it is the tar that is the bonding agent. Asphalt is usually delivered in blocks made up from a mixture of asphalt and sand. These blocks are melted on the site, mixed with chippings and laid hot. Concrete is now widely used for the

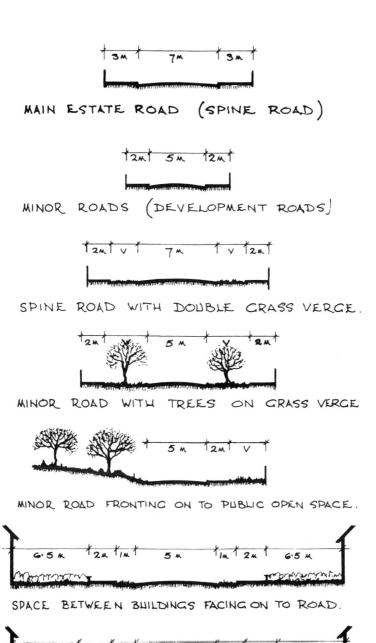

MAIN ESTATE ROAD (SPINE ROAD)

MINOR ROADS (DEVELOPMENT ROADS)

SPINE ROAD WITH DOUBLE GRASS VERGE.

MINOR ROAD WITH TREES ON GRASS VERGE

MINOR ROAD FRONTING ON TO PUBLIC OPEN SPACE.

SPACE BETWEEN BUILDINGS FACING ON TO ROAD.

SPACE BETWEEN BUILDINGS WITH TREES.

Figure 7.1   Roads in residential areas

57

surface of a road and it can be quite satisfactory, provided there is no need to cut trenches into the surface to reach pipes underneath.

(2) *Base course*   this is a course forming a cushion to help absorb the imposed load and to provide a fixing for the wearing surface. It can be constructed of lean-mix concrete, wet-mix macadam or dry-bound macadam.

(3) *Sub-base*   this layer lies between the foundations and the moulded earth below. It is primarily for purposes of drainage and where drainage is not needed it is sometimes dispensed with and the foundations are placed directly on the formation. It can consist of natural gravel, crushed rock or crushed concrete.

(4) *Formation*   this is the name given to the surface of the sub-soil which has been moulded with earth-moving equipment to form the lowest layer of the road; it is known as the sub-grade.

The Footway   The traditional material for footpaths of all sorts was natural paving stones, laid on a bed of sand with hardcore foundations. Most paving stones are now man-made and concrete is often laid *in situ*. Mastic asphalt is becoming more commonly used, sometimes by itself, and sometimes for that part of the pavement between the kerb and a double row of slabs.

The Grass Verge   This is often left to chance, but in a well-planned development sufficient top soil should be laid and graded and seeded down with grass. If quick results are required or the work has to be carried out in the nonplanting season, the surface can be turfed. This is generally a more expensive way of doing the job and it demands more attention until the turf is properly established. If trees are planted on the verges the top soil should be sufficiently deep to enable the roots to establish themselves. Services are generally located under grass verges. There should be no obstruction such as trees or street furniture close to the edge of the carriageway.

Road patterns appropriate to residential developments, town centres and industrial estates are considered in the chapters devoted to each of these subjects.

REFERENCES

1. Colin Buchanan, *Traffic in Towns* (HMSO, 1963)
2. Ministry of Transport, *Layout of Roads in Urban Areas* (HMSO, London, 1966)
3. Department of the Environment, *Transport Design Bulletin, Residential Roads and Footpaths* (DOE, London, 1977)

8 RESIDENTIAL DEVELOPMENT

This book is written at a time when the attitude of both central and local government towards the provision of housing is undergoing a process of change. The building of new towns is slowing down and the once familiar sight of widespread clearance of decayed housing to make way for whole-sale redevelopment is being replaced by schemes of improvement and rehabilitation.

These changes are taking place in a period of economic stagnation when little private development is taking place and when there is a general lack of confidence in the building industry. Demand in the private sector is increasing but developers still seem to be holding back. The revival of demand will cause values to rise but the Government will bring pressure to bear on building societies to restrict lending in an effort to hold down prices. Building costs have risen sharply and are continuing to rise. Interest rates are not stable for very long and this causes cash flow problems.

Under normal conditions residential development is carried out by or on behalf of many different development agencies and the size and type of development can vary over a wide range. (The agencies are considered in some detail in chapter 2.) The present chapter deals with small and medium-size estates in some detail and takes a general look at housing in new and expanding towns. It also indicates the variations in design criteria to meet the requirements of particular clients. The factors taken into account when designing a residential estate for sale freehold will be different from those where the client is a local housing authority, a new town corporation or a housing association. Whatever the development agency, planning approval must be obtained and the statutory requirements outlined in chapter 4 will have to be met. Of the planning standards already referred to, residential density is probably the most important and certainly the most controversial and, as such, requires examination in some detail.

RESIDENTIAL DENSITY

This is a system of relating the amount of accommodation to the size of the plot of land it stands on. It is expressed as the number of habitable rooms or persons to the hectare. (The hectare, $10\,000\ m^2$, unit symbol ha, will slowly replace the acre over the next few years. The conversion from acre to hectare for this purpose can usually be 1 hectare = 2.5 acres, except in large schemes.) The term 'bedspace' is now being used in place of 'habitable room' or 'person' and this tendency is likely to continue. It is more realistic in terms of the actual accommodation than either of the other two terms.

Variations of this density formula are used for many purposes, but that which a developer will be most concerned with is 'net accommodation density'. This is the basis for calculating the number of habitable rooms the planning authority is likely to approve in a given area. This information is likely to be found in the development plan.

There is no statutory definition of 'habitable room' in planning law but the expression is generally used to describe a room between 9 m² and 18 m² in area used for living or sleeping purposes. A half-room is between 4.5 m² and 9 m² in area.

When population density (persons to the hectare or acre) is given it can be converted to accommodation density by applying the occupancy rate which, in a large number of planning authority areas, is one-to-one so that the number remains the same. To calculate the number of habitable rooms on a given plot of land, the exact site area must be measured and to this can be added half the width of adjoining roads up to a maximum of 6 m. It is usual to include the whole of the road if a scheme spans the road and thus has both frontages. It is often possible to deduce from the density given the type of development that the planning authority has in mind.

It was never intended that density control should be mathematically applied, as the following extract from Planning Bulletin 2 shows[1]

'If precise density figures are applied too rigidly in controlling development the result might well be to produce uniformity and create unnecessary obstacles to the more intensive use of land.'

Under the sub-heading 'Control in practice' it goes on to say

'In development control what is important is not to apply some predetermined density control but to consider proposals in relation to the particular site and its surroundings, the proposed layout and the dwelling types.'

The bulletin then gives the following additional factors, which the planning authority should take into account in relation to density

> Light and air
> Access car parking provisions and waiting space
> Quality of design and layout
> Relationship to neighbourhood property.

In the summary at the end it suggests that higher densities present a challenge to architects, developers and local authorities and goes on to say

'They must ensure that the results are good to live in, good to look at and good neighbours to adjacent property.'

SPACE BETWEEN BUILDINGS

A distance of 21 m has long been accepted as the minimum to aim at when planning the layout of individual low-rise housing, although this standard is not always maintained. In the case of housing facing on to a road, distances are governed by the building line. This is a line prescribed at a given distance from the centre line of the road or from the edge of the road. The distance is usually related to the width of the road and the residential density of the area. In developments at the lower densities [around 100 habitable rooms per hectare (hrh)] the building line is likely to be 7.5 m from the edge of the road. (The word 'road' in this context means the carriageway, the footways and any grass verges.) This is reduced to 6 m and to a possible 4.5 m as densities increase. Permanent buildings may not be constructed in front of the building line but if all the buildings are along the same line the design can become monotonous. To give a long straight frontage a more interesting appearance it may be possible to set back groups of houses but this should be done boldly and with fairly large groups to achieve the best effect.

DAYLIGHTING STANDARDS

A useful test when considering daylighting is given in Planning Bulletin 2.[1] This test is to project a site line at an angle of 25° from the ground level at the face of the building and see that this line is not broken by other buildings. This and the following alternative method are illustrated in figure 8.1. An earlier recommendation in the Dudley Report (1944)[2] adopted an angle of 15° projecting from a table height of 3 ft 3 in (1 m) at a point 15 ft (4.5 m) into the building. Standards of daylighting for taller buildings are maintained by the use of daylighting indicators. They are used for both residential and nonresidential buildings so that they meet the British Standard Code of Practice (CP 3). The standard for housing is higher than that for other purposes. The indicators are in sets of four with one set designed for use from building to building and these are illustrated in figure 8.2. There is another set for use from the boundary or the centre of the road to the building. The indicators are a series of diagrams drawn on transparent material to the scale of the plan to be tested. Each diagram consists of two radial lines at a given angle with the space between graduated at 10 ft (3 m) intervals to show the permitted height of a building. The complete set of sixteen and details of their mathematical construction and how to use them are given in Planning Bulletin 5.[3]

SUNLIGHTING STANDARDS

The British Standard Code of Practice dealing with sunlighting recommends that for houses and flats the living rooms and, where practicable, kitchens and bedrooms also, should have the windows providing the main source of light, placed so that sunlight can enter for at least 1 hour each day during not less than 10 months of the year. An indicator for this purpose is shown in figure 8.3 and details of its construction and use are also given in Planning Bulletin 5.[3]

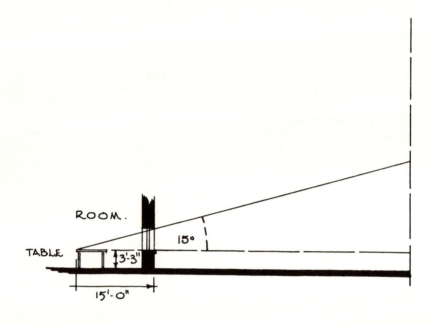

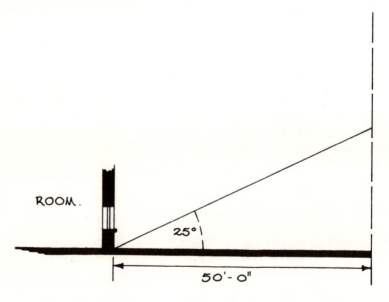

Figure 8.1   Daylighting angles

FROM OTHER BUILDINGS ON THE PLOT

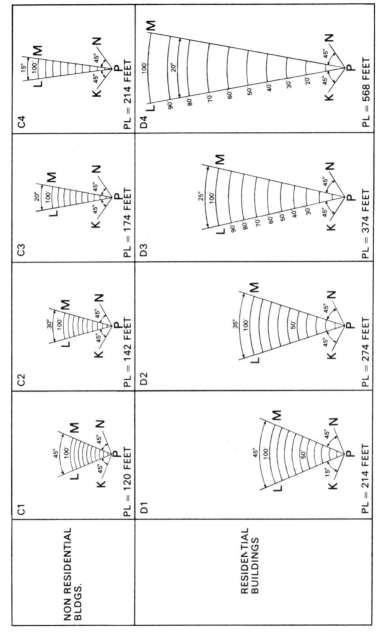

Figure 8.2 Daylighting indicators

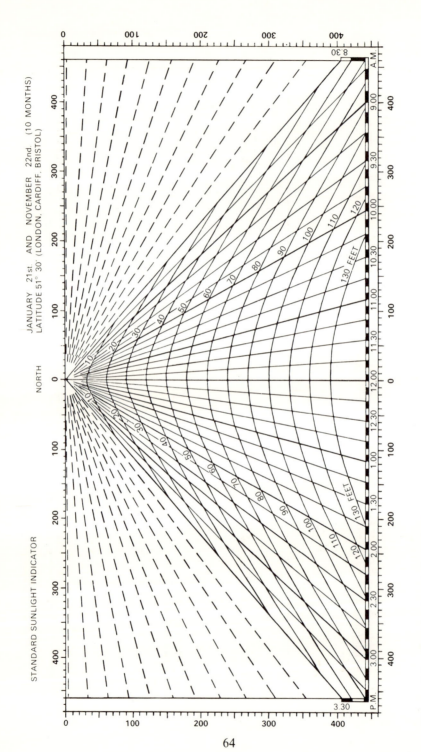

Figure 8.3   Standard sunlight indicators

64

Orientation is often difficult in residential development at other than low densities but when possible a house should be so sited as to allow sunlight into living rooms in the afternoon and the kitchen and bedrooms in the morning. See figure 8.4.

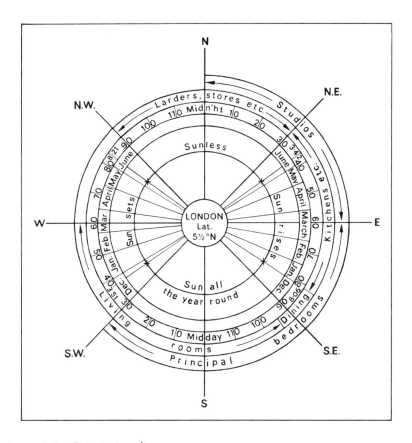

Figure 8.4   Orientation diagram

RESIDENTIAL ESTATES

Large-scale local authority development may include a mixture of both high and low-density housing in one estate, and an appraisal of such development is given in chapter 9.

This chapter takes each of the usual dwelling types in turn and suggests suitable densities and other design factors. It makes particular reference to the development agency most likely to carry out a particular type of development.

Detached Houses

This is usually a private development and the houses are built for sale freehold. Living in a detached house is, to many people, a status symbol; the house should therefore stand on a plot of land large enough to provide a garden of fair size and if possible garaging for two cars and also additional parking within the curtilage. To obtain a reasonable level of privacy the maximum density for detached house development is generally accepted as 100 hrh. If each house is taken to consist of 5 habitable rooms it means that there are 20 houses to the hectare. Single house plots in a rural setting are often very much larger but this type of development is outside the scope of this book.

At a density of 100 hrh the average size of an individual plot *including half the width of the adjoining road*, and after allowing 5 per cent for incidental demands on space, can be up to about 475 m². The building line is likely to be 7.5 m. With a rectangular site the frontage should be not less than 12 m to provide for a single garage on the same building line and adjoining the house. This allows for a plot depth of about 40 m but this must include half the road. In order to provide a double garage the width of the plot must be increased to not less than 14 m and this reduces the overall depth of the plot to about 34 m.

The frontage on a convex bend will need to be longer and on a concave bend it can be shorter but these are likely to compensate for each other if both sides of the road are within the same scheme. In both these cases the measurements should be taken at the building line.

Frontages can be reduced if garages are sited to the rear of the house but this usually requires a longer drive; it also reduces the size of the back garden unless the depth of the plot is increased proportionately. This results in little saving in land so that there is no point in adopting this less attractive layout unless peculiarities of the site demand it. The same reduction in frontage can be attained by siting the garages at the rear of the site but this will necessitate a rear service road, which will add to constructors' costs and show little or no saving in land. A small but interesting development of 8 houses on a plot of about 0.6 ha with the option of an addition of 3 on a small plot adjoining is illustrated in figure 8.5.

It is not usual to provide for open amenity space in this type of development, particularly on a freehold estate. Although it may look attractive if well kept there are inevitable problems of maintenance and upkeep. Attempts are sometimes made to form residents' associations for this purpose but they are seldom a success after the first flush of enthusiasm.

An interesting attempt to overcome this problem is being introduced at the new village of Martlesham Heath, Suffolk. (Comments on this development are given in chapter 9.) A company is formed and each house purchaser becomes a member (shareholder) of the company. The freehold of the open spaces is vested in the company, which in turn has the responsibility for their management and the power to levy annual charges on all the owner-occupiers to meet the costs. It will be interesting to see if this works satisfactorily.

To give a more spacious look to an estate, grass verges between the carriageway and the footways can be made wider than usual and planted with flowering trees. These can then usually be handed over to the local authority

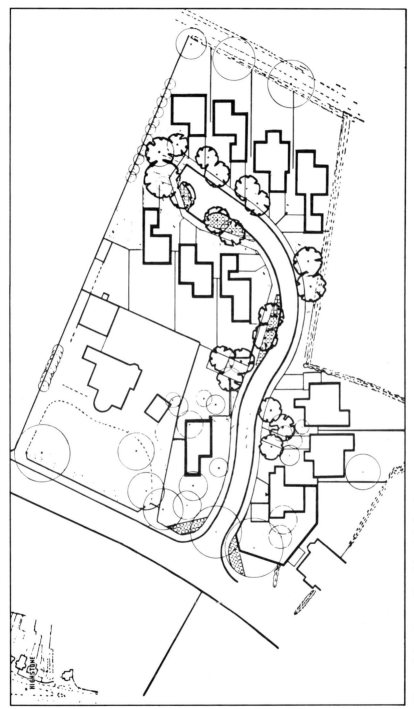

Figure 8.5 A small residential development

when they take over responsibility for the roads. Another way of achieving a similar effect is to plant grass and trees between the footways and the front fence of each plot and this will usually be looked after by the frontager.

Semi-detached Houses

This type of development usually affords the occupant about half the privacy possible in detached house development. For this and other reasons it is often treated as the poor relation but it should be remembered that it is still by far the most popular form of housing for the great majority of people. Much of this type of development is private and the houses are for sale freehold but local authorities still build 'semis', particularly in rural areas and they are also to be found in new and expanding towns.

For many years 30 houses to the hectare (12 houses to the acre) has been the generally accepted density and this still applies in many cases. A typical semi-detached house has $4\frac{1}{2}$ habitable rooms, which makes for a density of 135 hrh. Most urban planning authorities are prepared to permit a small increase in this number to 150 hrh on a suitably shaped site.

There should be at least one garage to each house and in private development this should be within the curtilage. In local authority or new town development it is often better to site the garages away from the house. The same number will have to be provided but by this means a garage is not allotted exclusively to the tenant of a house, as it would be if sited within the curtilage. Individual plots will have to be smaller to allow sufficient land for the garages. The building line is likely to be 7.5 to 6 m. With a rectangular site the frontage should be not less than 9.5 m to allow for a single garage adjoining the house and on the same building line. This will allow sufficient room between the garage wall and the boundary fence for it to be possible to walk round to the back of the house.

If the above density of 150 hrh is interpreted as 34 houses to the hectare and an allowance of 5 per cent is made, the average size of an individual plot can be about 280 m² and the maximum depth about 29.5 m from the centre of the road.

When considering the layout of a site for this type of development it is advisable to think in terms of double plots rather than single ones. In the case of a restricted frontage it may be possible to set the garages back to the rear of the houses but this is not very satisfactory. Since it usually means dispensing with the boundary fence between pairs of houses, the plots must be proportionately longer if there is to be a worthwhile garden at the rear. The same point applies if the garages are sited at the end of the garden, when a service road has to be provided for access. In a private development, open amenity space should be treated in the same way as for detached houses. Where a local authority or a new town is the development agency the problems of long-term maintenance and upkeep do not exist because in both cases there is a department with staff and equipment for the purpose.

Terraced Housing

A number of different classes and types of house are built in terraces but in this case the name is given to houses built in rows by local authorities or new town corporations. Garages are provided but usually sited in groups. They are available to the tenants of the houses but not strictly allocated to them and in this way there is flexibility in the letting policy.

Densities of up to 200 hrh are possible and some very narrow frontages have been achieved, but for the purpose of this book 190 hrh and frontages of 6 m are used. Houses are assumed to have $4\frac{1}{2}$ habitable rooms, which is about 42 houses to the hectare. Allowing about 32 m for the overall depth of the garden measured from the centre of the road, the land actually allocated to each house will be 180 m^2 and at 42 to the hectare only about 80 per cent of the land will be devoted to house plots. Much of the remaining land will be required for service roads, garages and parking, and this can be calculated on the basis of a minimum of 32 m^2 per car.

This type of development is probably the most suitable for Radburn layouts, as recommended in Design Bulletin 10.[4] Radburn is the name of a 1930s development in New Jersey, USA, which introduced a new concept in neighbourhood planning. It was based on the principle that vehicular access to a house plot should be off culs-de-sac pointing like fingers into large areas planted with grass and trees. Footpaths off these areas should be the means of pedestrian approach to the houses. Hence a substantial degree of vehicular segregation is attainable. The internal layout of the houses should be designed so that the living rooms look out on to grass and trees. Figure 8.6 illustrates the original Radburn layout.

The main difficulty experienced in the United Kingdom in adopting this technique to meet local conditions is one of density. The original US development was at about 90 hrh whereas in the United Kingdom we think in terms of up to 175 hrh. This means much less open space and the loss of its most valuable feature: that of a flowing landscaped area of open space, penetrated for short distances by minor roads serving houses looking on to a miniature private park.

Town Houses

This is the name given to houses that are usually on three floors and built in terraces, with the garage the main feature of the ground floor. The town house is likely to be part of a private development and densities similar to those for normal terraced housing can be achieved. The main living area is at first-floor level and this is often open-planned. The width of the frontage depends on the number and size of the bedrooms: 6 m will just allow for three double bedrooms on the second floor, making each house the equivalent of 5 habitable rooms or more. The depth of individual plots measured from the centre of the road should be deeper than the normal terraced house to allow for a secondary service road to run parallel to the existing road, since this is a usual planning requirement when there is vehicular access every 6 m of frontage (see figure 8.7). As with other forms of private development it is not

69

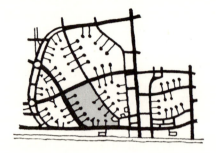

Figure 8.6 Radburn, New Jersey (bottom) one of the completed neigh-bourhoods, (top) the block plan

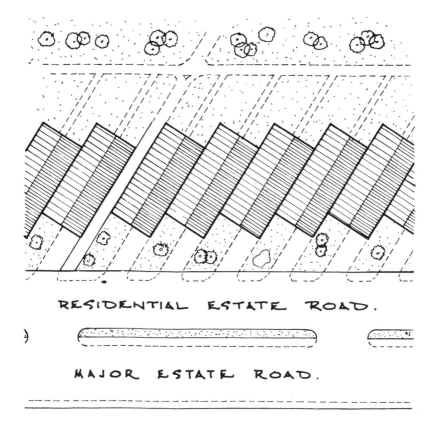

RESIDENTIAL ESTATE ROAD.

MAJOR ESTATE ROAD.

Figure 8.7   Town houses built in echelon

usual to provide open amenity space but there is likely to be an opportunity
for tree planting and the provision of shrubs on the island formed between the
service road and the original footway, as also shown in figure 8.7. It is not
possible to do much with the frontages of individual houses since almost half
is taken up by the hard surface leading to the garage. The most satisfactory
treatment is to plant low-growing shrubs and have just a low wall at the front
of the garden but with no wall or fence between house plots.

   Orientation is difficult at these comparatively high densities, particularly in
this terrace type of design but it is possible to face houses at an angle to the
road by building them in echelon formation. Construction costs are higher,
particularly in the roof structure, but the pleasing effect obtained may well
justify the extra expenditure.

Patio Houses

This name originated in estate agents' sales literature and has now become the

71

accepted name for a house enclosing an inner courtyard rather than having the conventional back garden. It can be a local authority or private development, but in a local authority development it might be desirable to site the garages in groups as illustrated earlier and not have them within the curtilage and forming part of the enclosure. In new town development patio houses can be designed with or without contiguous garages, depending on location and the type of occupiers the houses are planned for.

Figure 8.8 shows plans for two typical patio housing designs, one with front access and the other with the garage off a service road at the rear. Densities are similar to town house development but, owing to the size of the actual house plot, there are often small areas of land, footpath and private service roads which have to be maintained. As stated earlier these are no problem for local authorities or in new towns, but may ruin an otherwise good private development unless proper provision can be made for their care and maintenance. There is also the question of ownership when the developer moves out after all the houses are sold.

Another type of development which can be treated as a variation of patio housing is courtyard housing, illustrated in figure 8.9. This scheme was chosen by Edinburgh University as the subject of a sociological study which included views on the privacy afforded by this form of development. It was a local authority development for Dundee Corporation and part of the conclusion drawn from the research indicated that most of the tenants experienced little difficulty in accepting it as a form of housing. In this particular development additional facilities were provided which had to be shared and this was where difficulties arose.

The uniformity in the design of the Dundee courtyard housing scheme is in direct contrast to the design criteria contained in the Essex design guide. This was probably the best known publication of its kind issued by a planning authority.

Essex planners were disturbed about the proliferation of sprawling and monotonous speculative estates which they attributed not only to the profit motive but also to the inflexible standards then being enforced. They were able to reach agreement with the county engineers to relax certain standards. (This relaxation is now taking place at national level with the publication of a bulletin referred to in chapter 7.) It enables the developer to reduce the large areas of a residential estate devoted to roads and footpaths. The planner is able to accept and even encourage higher densities, thus helping the developer to set off the higher material and structural costs the design guide advocates to produce an interesting streetscape and a sense of space.

The guide identifies the Essex vernacular features, such as dark timber boarding, red bricks and steep pitched roofs. Other distinguishing features of the guide are gently curved streets and village-like layouts, achieved through the use of mews courts and by bringing the houses directly up to the footpaths.

Low-rise Flats and Maisonettes

This is a comparatively modern form of development and it is an attempt by local authorities to obtain higher densities without resorting to high-rise flats

VEHICULAR ACCESS AT REAR.

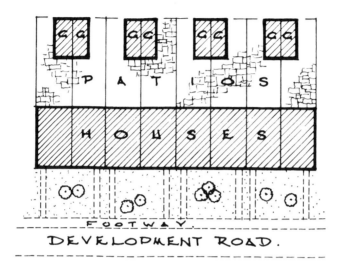

VEHICULAR ACCESS AT FRONT.

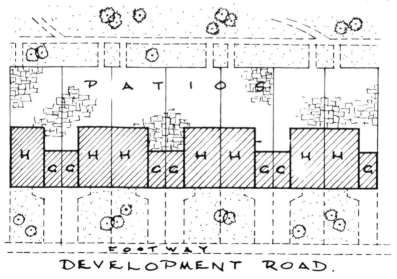

Figure 8.8   Patio housing

73

which, for a number of reasons, have become unpopular with local authority tenants. They are built in slabs and are usually restricted to five storeys, and no lifts are provided.

Figure 8.9   Courtyard housing. Dwelling sizes are indicated: for example, 4/3 means four-person, three-room

The ground, first and second floors can be flats but the two top floors should be maisonettes with living rooms at third-floor level and internal staircases to bedrooms on the fourth floor. By this means there are only three flights of common user stairs. Densities greater than 250 hrh can be attained, making this form of development useful in heavily populated urban areas, particularly in schemes of urban renewal following slum clearance. The provision of open amenity space and other landscaping is most important in this type of development since at least 75 per cent of the residents will be without their own gardens. This figure may be 100 per cent if it is the policy to have no private gardens even at ground-floor level.

In flat development all garages and parking facilities should be contained within the scheme since it is seldom possible to overflow into adjoining streets. The space around blocks of flats required to meet daylighting standards generally provides sufficient space, but careful siting of garages and planning of hard standings for parking is essential.

Residents should be able to leave their cars close to the block they live in and visitors' parking should be easy to find and conveniently situated. Areas for parking should be marked out and if possible allocated to individuals or to classes of vehicle. What could otherwise be a pleasing and well-planned estate can look like a used-car sale lot or even a car breaker's yard if parking is not controlled. Earth moulding and the planting of evergreen shrubs can be used to demarcate the area allotted to parking. The minimum recommended internal width for a garage for a car of standard design is 2.4 m. The door should preferably be the full width but it may be slightly smaller. The minimum internal length is 4.8 m. The minimum depth of the forecourt should also be 4.8 m. The space between garages facing each other should be not less than 7.2 m. If on one side only this may be reduced to 6.6 m.

Smaller dimensions can be used for open parking areas and figure 8.10 illustrates how areas of different size and shape can be used to the best advantage. These illustrations and other details, including those of multi-storey garages, can be found in Design Bulletin 12.[5]

High-rise Flats and Maisonettes

This is not a popular form of local authority development, but some are still being built. There is no other satisfactory way of obtaining densities of around 500 hrh and at the same time providing amenity areas and parking facilities. Planting and landscaping become more important, as does the provision of additional facilities. High-rise flats are more acceptable when they are part of a mixed development. This principle has been adopted in Birmingham, where tower blocks form the centre-piece, surrounded by two-storey houses facing on to the open areas between, thus creating a much adapted mini-Radburn layout. The problems of fire protection and the means of escape become increasingly important as the blocks rise higher.

Communal Facilities

In a large estate of local authority housing, particularly one which includes flats, additional facilities are reqired. In a low-rise housing development these would normally be provided within the curtilage of each dwelling.

Laundry and Drying Rooms   With the increase in the ownership of washing machines and spin driers there is now much less need for a communal laundry,

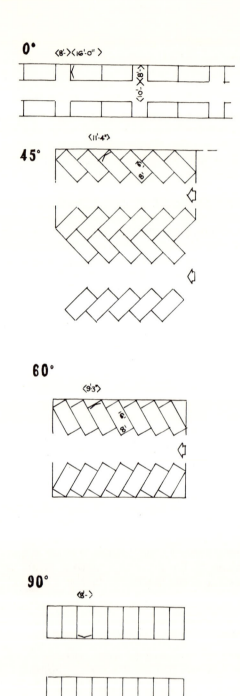

Figure 8.10   Car parking

but an outside drying ground is still considered worth while. This should be so sited as to be screened from the public thoroughfare but visible from the dwellings.

Community Centre   This can provide a meeting place for a residents' association, and should be available to the over-60s and other organisations. It might house a branch library, a clinic and a consulting room for visiting doctors. It should have some accommodation set aside and kept warm and comfortable for people to just meet. It may also include a hobby room. It should not be considered as a substitute for any of the established institutions such as the pub, the working men's club, the church hall or the youth club but as complementary to them.

Amenity Spaces   People living in flats will not have a garden to sit in nor will children have their own yard or garden to play in. These must therefore be provided on a communal basis. It means that the space between blocks required to meet daylighting standards should be used to the full after such essentials as garaging and parking have been provided for.

Children's Play Areas   The play facilities for children should be provided according to their age.
   Toddlers   An area of about 50 m² should be enclosed with 0.6 m chain-link fences, with a self-closing gate. The surface should be part green and part asphalt, or other nonabrasive paving material. It should be sited so that it can be seen from the flats, and should be fitted with climbing poles and crawling tunnels.
   Fitted playground   For the next age group a playground should be fitted with swings and other equipment. They should be robust, vandal-resistant and as safe as possible. The fence should be 1.35 m high with a lockable gate.
   Playground for ball games   This should be open to children of 14 and above. It should have a high fence with a lockable gate and be sited well away from the dwellings so that the children can be noisy without disturbing other people.
   Adventure playground   A suitable site enclosed by a high fence and with a lockable gate should be provided, but the equipment and the running of it should be left to organisations with playgroup leaders.

TV and Radio Masts   These are considerable assets in a modern development because they take the place of individual aerials, which can ruin the roofscape of an otherwise interesting group of buildings. A further advantage with masts is that where there are no chimney stacks the problem of fixing is eliminated.

Car Washing   This should be provided in the vicinity of the garages but should be available to all residents on the estate.

Housing for the Elderly

As society and family structure change, the demand for small but partially independent housing will grow. The following quotation, from an internal Department of the Environment paper by Derek Fox, adequately sums up the social approach needed in planning for the elderly

'The construction of suitable accommodation for old people is not simply dependent on analysing physical needs, calculating the scale of the provisions and assembling in economic fashion an aggregate of the parts. It depends much more on learning how old people live, how they interact, what they want to make of their lives and then translating the knowledge into an effective and sympathetic physical solution'.

This type of development is normally made by a local authority, a new town or a housing association. Charitable organisations used to build almshouses, but this seldom happens today except as part of a replacement programme. There are also many schemes for the conversion of large houses into nursing homes or 'guest houses for elderly gentlefolk', but these are outside the scope of this book. Mandatory standards are laid down by central government in respect of space, design, heating and fittings. There are other standards which are not mandatory but are encouraged—these are related to communal amenities and a warden service. The traditional picture of retirement to a cottage in the country has influenced many authorities towards bungalows with individual gardens. This may still be feasible in rural areas but in towns the high cost of land and the demand for residential sites will force most urban authorities to provide flats.

The amount of space and the internal layout of the accommodation provided vary with different authorities. Bedsitters were once the most common form of development but most housing managers are now of the opinion that people prefer a separate bedroom, even if this means additional heating costs.

A case has been made for at least some of the dwellings to have a second bedroom but experience has shown that these are often unoccupied for a large part of the time. Views are changing on whether or not housing for the elderly should form part of a mixed development and the general feeling now is that there should be partial segregation. Elderly people should be able to see what is going on around them but should be able to avoid taking part if that is what they want. It has been considered for some time that 30 dwellings should be the normal maximum in one development to achieve this limited involvement but other design considerations may justify an increase in this number on certain sites.

Housing for Single People

This type of accommodation used to be provided by the conversion of old houses into bedsitters and by purpose-built hostels run by such organisations as the YMCA/YWCA. There is now a growing demand for dwelling units designed to meet the needs of the single person. For the purpose of considering design criteria single people are classified into two groups

78

(1) the older working single person

(2) the young more mobile person.

A study of the special requirements of single people is contained in Design Bulletin 29 which also contains the description of a scheme at Leicester.[6]

Housing Mix

There is something to be said for councils providing different-sized units in one estate, particularly if the housing list shows that a number of different-sized dwellings are required. When different-sized units are provided in one locality, tenants can be moved into larger dwellings as their children grow and into smaller ones as their children leave home. This can be accomplished without their having to leave the district and move away from their friends. It is an excellent arrangement in theory but one which does not really work well in practice. People are generally reluctant to move out of a family house even if only the husband and wife are left, and the house is much under-used. New towns must try to attract a cross-section of people to build up a balanced community. They must provide houses of all sizes to rent in addition to making land available for private development. This does not mean that each estate should contain all types and sizes of house. As a general rule people like to live among people who are in similar circumstances to themselves and with the same sort of background. Private development must be governed by the profit motive, which will usually influence the type, class, size and price range of the houses in a proposed scheme. Although these considerations are vitally important in private development they are subject to the control over development exercised under planning law. Unfortunately this often means that the residential density which the planning authority is prepared to approve will go a long way towards dictating the type of development that takes place.

A bold attempt to break away from the conventional approach to a housing mix is outlined in the report on the new village at Martlesham Heath, Suffolk, in chapter 9. In the layout of the first phase of this development, houses of different types, sizes and values are deliberately brought together to form pleasing village-like groups. The motive in this case is aesthetic and possibly also social rather than economic.

Starter Homes

The Government is keen to promote the development of housing units suitable for people who do not live in average families. A circular[7] was published in 1975 with the object of speeding up the supply of dwellings and improving the housing mix. It points out that more than half the households in the United Kingdom consist of only one or two people. Starter homes and other forms of low-priced housing are encouraged.

There are also proposals for infilling in existing council estates to increase densities and also suggestions for the reintroduction of something like the 'prefabs' of the immediate post-Second World War period, and the use of mobile homes on cleared sites awaiting a more permanent development.

REFERENCES

1. Ministry of Housing and Local Government, *Planning Bulletin No. 2, Densities in Residential Areas* (HMSO, 1952)
2. Ministry of Housing and Local Government, *Design of Dwellings* (HMSO, 1944)
3. Ministry of Housing and Local Government, *Planning Bulletin No. 5, Planning for Daylight and Sunlight* (HMSO, 1964)
4. Ministry of Housing and Local Government, *Design Bulletin No. 10, Cars in Housing: 1* (HMSO, 1966)
5. Ministry of Housing and Local Government, *Design Bulletin No. 12, Cars in Housing: 2* (HMSO, 1967)
6. Department of the Environment, *Design Bulletin No. 29, Housing Single People: 2* (HMSO, 1974)
7. Department of the Environment, *Circular 24/75* (HMSO, 1975)

9 COMMUNITY DEVELOPMENT

To create a community rather than just to build streets of houses, shops and workplaces has long been the ambition of many people. In the United Kingdom we have the industrial village, the garden city and the post-Second World War building of new towns. All these illustrate in their own particular way attempts to satisfy that ambition. Much has been written about new towns and many readers will be familiar with at least one. Most of them are worthy of detailed study.

More recently attempts have been made by the private sector to create communities on a village pattern. One of the latest developments of this kind is taking place at Martlesham Heath near Ipswich, Suffolk. Building work began in 1974 and already a semblance of a village can be seen taking shape.

This development does not conform to some of the more conventional guidelines suggested in earlier chapters. It has a number of the features contained in the Essex design guide and this makes it an interesting project to examine. The remainder of this chapter takes the form of a report on the development.

MARTLESHAM HEATH VILLAGE

The Developer

The development is being carried out by Bradford Property Trust Ltd, which owns the freehold of the site together with other land in the immediate vicinity, including an industrial estate to the east. All the houses are to be sold freehold but there are a number of long-lease maisonettes and flats. A management company has been formed to take over the freeholds. The company will also take over the village green, other open space and any land within the development area not conveyed to individual purchasers or not vested in the local authority. Each purchaser becomes a member (shareholder) of the company and is required to make an initial payment of £20. It is hoped by this means to overcome the problems usually associated with the care of open spaces.

The Site

The village is being built on a site situated a little to the south of the junction of the A1093 with the A12, close to the village of Martlesham and about 5 miles east of Ipswich.

The proposed layout of the whole site is shown in figure 9.1; the village is made up of 12 clusters of houses which are called hamlets. Some are large with up to 200 dwellings and others are quite small by comparison, with as few as 30.

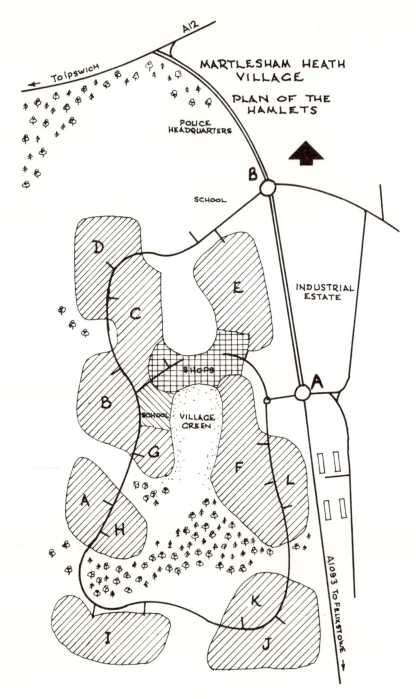

Figure 9.1   Martlesham Heath Village

82

The Locality

The locality is generally fairly flat and typical of a disused airfield, of which there are many in East Anglia. There is a wooded area, some scrub land and a number of hangars, barrack blocks and other permanent buildings around. Most of these blend into the landscape reasonably well. The one exception is a group of buildings that includes the old control tower. This is still a prominent feature, standing on land to the east of the development site and south of the main industrial area. It is at present occupied by the GPO as a research station. The land between the A12 and the development site is occupied as a police headquarters. There is a small local flying club in temporary occupation of land to the west but the tenancy can be terminated at short notice. The A1093 forms the boundary to the east and also serves as a positive barrier between the village and the industrial estate.

Access and Circulation

The present access to the site is off the A1093 at the second roundabout, shown as point A on Plan 1. When the development is more advanced the second access will be opened at the first roundabout, point B.

The main estate road forms a 'figure of eight' but there may be some changes where it joins the village centre when detailed plans for that area are finalised. The development roads are all culs-de-sac, as illustrated in figure 9.2 which shows the detailed site layout of part of hamlet F.

The pavements alongside the carriageway are linked to an independent pattern of footways leading from the houses to the centre of the village. A number of footbridges are planned to cross the main traffic routes.

House Types

In the first phase (hamlet F) there are ten different basic designs. Each of these can be varied in a number of ways by using different kinds of brick or by cladding the roof with a different kind of slate or tile. Houses that are rendered can be painted in different colours and there are several different designs for the porch at the front door. There is very little grouping of particular house types except in the case of the maisonettes and flats and there are no continuous terraces of one type of house. The houses are sited in informal groups with some structurally linked, but they are seldom all on the same frontage line. No building line, as such, seems to have been prescribed. Particular attention has been paid to the houses fronting on to the village green.

The house types vary from detached houses with five bedrooms to one-bedroom flats. The internal circulation is in most cases well planned and maximum use is made of the available space.

Garages and Parking

Most of the houses have garages within the curtilage. Some form part of the

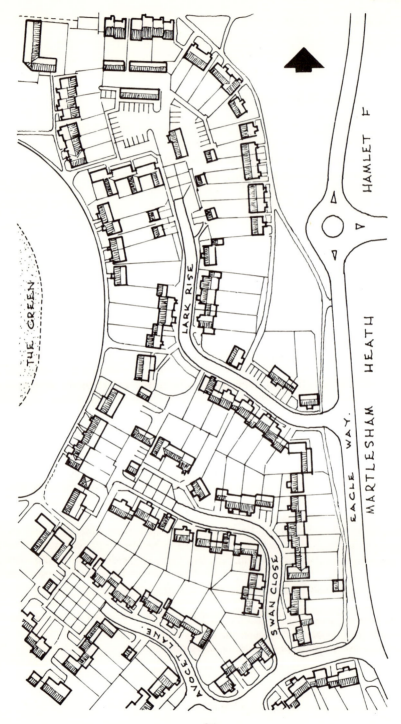

Figure 9.2 Hamlet F, Martlesham Heath Village

structure and some of the larger houses have double garages. Some of the maisonettes and flats have detached garages which are included in the purchase price and some flats have garages available which can be purchased, but these are not strictly allocated to a particular flat. Parking for visitors is available at the head of most of the culs-de-sac and provision for parking is made throughout the residential area. There is also a large car park associated with the village centre.

Ancillary Uses

The village green, which is intended to be the central feature of the whole development, has been laid out and grassed down and is already quite well tended. The developers claim that it is large enough for county cricket. Most of the other ancillary uses are still only at the planning stage but land has been allocated for the following.

(1) *Supermarket and other shops* The final decision on the size of the supermarket and the number of other shops will not be made until the result of a pending planning appeal is announced. This appeal is being made by a firm wishing to build a hypermarket quite close to the development site. If the appeal is upheld and this large store is built it must have the effect of reducing the number of people living in the village who would use the village store and other shops.

(2) *Public house*

(3) *Coffee house*

(4) *Offices* No details are given but the developers intimate that they would be willing to consider meeting any specific requirement for either professional or commercial offices.

(5) *Banks*

(6) *Community hall* This would probably house the offices of the management company.

(7) *Squash courts*

(8) *Two schools* The first school is to be built close to the village centre.

(9) *Amenity areas* The most important of these is the village green and this has already been mentioned. This is only part of about 20 hectares which are planned for amenity use. A large area of open heath and woodland between F, H and K hamlets is to be preserved for all time. There are small patches of green between all the hamlets and also some quite pleasing small grassed areas with trees and shrubs along the footways, described earlier. A prime site close to the village centre has been reserved for old people's accommodation.

(10) *Landscaping* The village green has been grassed down and a number of trees have been planted around the edge. These will doubtless be added to as time goes on. The heath and the woodland already have a look of maturity and a natural charm. Some of the tree planting has been carried out and a number of undulating mounds have been formed to close certain unattractive vistas in the direction of the industrial estate and to form a sound barrier. There are plans to plant trees to screen the industrial estate from the village.

(11) *Services*   Both gas and electricity are laid on and all the houses and flats have small-bore, gas-fired central heating with a programmer. In addition, all the houses have one open fireplace. The electricity and telephone services are underground and there is also an underground service to TV and VHF aerials located on the nearby Post Office tower.

(12) *Employment*   A modern industrial estate within walking distance of the hamlet will provide work opportunities.

Conclusions

The controlled development of this village, although as yet in its early stages, has started well. There are indications on all sides that the developers are anxious to create something worth while. In their promotional literature they pose themselves a question, and their answer is worth quoting since it exposes the philosophy behind the planning of the development. In answer to the question, 'What is a village?', they state

> Groups of houses rather than identical dwellings
> A village store
> Shops and a pub overlooking the green
> Schools
> Open spaces
> Variety
> A community with pride in itself.

The effort put into the design and planning of this village to date deserves encouragement. If this effort does not flag as time goes on, it deserves success.

In the author's opinion, there is a slight danger that the desire for variety and individualism, if allowed a completely free rein, may result in producing a bitty mass of uncoordinated shapes, textures and colours rather than the village groupings that the developers seek. Developers are right in trying out those ideas in the largest hamlet first. It will be much easier in later phases, when dealing with smaller numbers, to apply the same grouping principles, or to make changes based on experience gained in the first phase. The design team know where they are going and to date they are on course. The scheme is an interesting one and shows all the signs of being worthy of a visit from time to time during its growth over the next few years.

10 SHOP DEVELOPMENT

The development agencies most likely to be concerned with shop development are new town corporations, local authorities engaged in town expansion or urban renewal, and the private developer.

LOCATION OF SHOPS

Tradition and long-standing shopping habits probably have more influence on the location of shops than any other factor. For many years the High Street has been the most important shopping area in the average town and in many towns the High Street is still the main road through the town. Some of the pressure has been relieved by the construction of a bypass but the heavy through traffic has been replaced by local cars. Notwithstanding the danger and, possibly, the inconvenience, many people are still attracted to the High Street shop and often ignore the safety and comfort of a nearby pedestrian-only precinct.

In a curious way people who are prepared to go along with the tide in most things seem to assert their right to freedom of choice when it comes to where they shop. In new towns there must be new local shops and new shopping centres and people accept this, but in established communities a decision to build shops away from the traditional localities may well be doomed to failure.

SIZE AND NUMBER OF SHOPS

The growth in the number of supermarkets and other bulk-purchase stores has introduced a new factor in the assessment of the appropriate number of shops for a known population. The provision of a shop is a two-way affair. The public feel entitled to demand a sufficient number of shops within reasonable distance of their home to meet their everyday needs. The shopkeeper has the right to expect a reasonable living in return for the capital outlay entailed and the labour he puts into the running of the shop. The shopping public cannot have it both ways—the more money they spend outside the immediate vicinity of their homes the fewer shops there will be in that vicinity. This is an economic fact of life; it may take some time to operate but it must do so in the long run. Shops only survive on the profits made from the goods or services they sell. A number of the major costs, such as rent and rates are of a constant nature and are not directly related to takings.

Another major factor that affects people's shopping habits is their mobility. The whole concept of out-of-town shopping is to cater for the person who is prepared to buy in larger quantities than they would otherwise do and carry their purchases home in their own car. There is a steady increase in this type of shopping but planning applications are in most cases resisted by planning authorities. There are no firm indications that the present trickle of consents will develop into a flood tide and switch the centre of gravity from the town centre to isolated groups of buildings in country areas, although this is thought by some people to be a distinct possibility. The substantial increase in the price of petrol and other motoring costs may well stabilise the motoring population at a little above the present level. This would mean that there would still be many people who are not able to shop by car.

The growing popularity of the deep-freeze cabinet is the third important factor to upset the balance in retail trading. There are traders, particularly butchers, who specialise in bulk sales of meat at reduced prices and who are prepared to cut whole carcases into joints ready for storage at home.

Notwithstanding the economic uncertainties and the possible decline in the population, the following trends should be taken into account. The total number of shops in the United Kingdom is declining at the rate of 1 per cent per year and this trend is likely to continue, and may well increase. Department stores are maintaining their 5 per cent of the total trade but independents are losing out to multiples at a steadily increasing rate.

Some interesting figures for the year 1974/5 produced by the Institute of Grocery Distribution show a considerable increase in the size of food stores. The average size of those closed was 125 m^2 and that of those opened in the same year was 850 m^2.

In the light of the wide variation in the size of shops required in the retail trade it may be easier to estimate need by relating floor space required to population. Most nonmetropolitan counties are likely to produce similar figures, but for the purpose of this exercise those for Suffolk are used. In 1975 there was a population of 545 800, and it was estimated that at the time there were 3 757 000 ft^2 of shopping floor space in the county. Stated in metric terms this means that there were approximately 0.633 m^2 of shopping space per head of population. For the purposes of this exercise, taking a shop of 100 m^2, there would be 1 shop for about every 160 people, but this seems high in the light of local experience.

To examine this figure in relation to shop takings, it is necessary to look first at local spending power. Taking the average (1975) amount spent on food for a household of four as £16, and using government statistics, published for 1975, that for every £100 spent on food £132 was spent on other goods, the takings of a hypothetically average shop of 100 m^2 would therefore be £77 200. It is fully appreciated that averages can be dangerous but, provided all local factors are taken into account, this can be a good starting point.

In the long term there may be substantial changes in the whole pattern of shopping but at the present time there are still indications of a demand for the conventional type of shop. A good rule could well be to build shops only where a real need can be shown.

TYPES OF SHOP

If shop development is taken in its widest sense it includes all forms of retail outlets except markets and mail-order warehouses. This means that a developer may be concerned with any of the following types of building used in the retail trade

Shops as part of a mainly residential development
A shopping precinct or neighbourhood centre or subcentre
A group of shops forming part of an urban renewal scheme
Shops forming part of a major town centre redevelopment scheme
Out-of-town shopping centres or hypermarkets
Cash-and-carry or discount warehouses.

The remainder of this chapter is devoted to considering the basic planning requirements to be taken into account in respect of each of the above.

The Single Shop

The corner shop still has its use today and when situated in the right place it can be of considerable service to people living around, and it can still be quite a lucrative business, notwithstanding the additional work entailed by the introduction of VAT.

When a single shop is being considered as part of a residential development, the size and density of the proposed development are not the only factors to be taken into account. On the one hand there may be shops already conveniently situated which meet all the requirements of the additional population in the area visualised by the new development; on the other hand there may be no shops in the immediate locality but a number of people already living there would be glad of a general shop within short walking distance. Numbers are always difficult and they are not the only criterion, but a fairly isolated community of 400 to 500 people may well support a single general shop.

The shop should not be sited right on a corner but should be close to a corner. It should have a pull-in off the carriageway and, if possible, rear access for delivery lorries. Although as a general rule a shopkeeper or manager does not like to live over the shop, there are still people who prefer to live on the premises of this sort of one-man business. If living accommodation is provided it should be interconnected with the shop but there should also be completely separate access to the flat above so that visitors do not have to pass through the shop. There may be a need to garage a trade vehicle in addition to the normal provision for a private car.

The Parade of Shops

The second type of shop development likely to be built as part of a mainly residential development will be the small parade of shops. Most of the same factors already considered apply to this type of development, except that there is less likely to be a demand for living accommodation over the shop, but this does not mean that flats over shops are not appropriate. It may well be a satisfactory way of increasing residential density without detracting from other features of the scheme.

A minimum of five shops in a terrace should be the next stage up from the single shop. People who come to a group of shops expect to be able to buy all they are likely to need on a mid-week shopping trip. If they find that such basic commodities are not available they are likely to go further afield and miss the local shops completely.

In order to be satisfied that there are sufficient potential customers in the immediate area who would be likely to use the shops, and since this would be on a somewhat casual basis, a population of 1800 or over would be needed to justify the development. With such a limited number of shops it is important to have the right occupiers. The local authority or a new town corporation could arrange this by putting the shops out to selected tender by suitable trades. A private developer would not normally be able to exercise the same sort of control but he should only sell or let to suitable businesses so that the retail coverage referred to earlier is complete as far as is practicable.

An appropriate size would be a frontage of about 5.5 m and a depth of 15 m. They should preferably be of cross-wall construction so that the shop front and the internal partitioning can be to the individual requirements of the occupier. The frontage should be set back enough to allow for a service road at the front and wide enough to allow a limited amount of parking. The pavement in front of the shops should be wider than the usual 2 m and there should be a service road at the rear for the delivery of goods. Figure 10.1 shows a typical parade of shops suitable for this purpose.

An interesting example of urban redevelopment in South London is shown in Figure 10.2. It consists of shops with storage and a double garage at semi-basement level, offices on the first and second floors and so-called patio flats on the top floor.

The Shopping Precinct

A precinct can be part of a neighbourhood centre in a new town. It could also be part of a town centre extension in an expanding town, part of a scheme of urban renewal in an established community or even the shopping centre in a much enlarged village.

There have been pedestrian shopping streets for many years in such places as Chester, York, Brighton and Tunbridge Wells. There are also many old-established enclosed arcades, such as the Burlington Arcade in London, but the modern concept of the shopping precinct was originated by Alker Tripp. When considering the role roads should play in the post-Second World War period, he was convinced that the days of the all-purpose road were numbered owing to the large increase in the number of cars likely to be on the roads. He recommended the construction of bypasses to keep cars out of the town centre but he also advocated the segregation of vehicles entirely from new and rebuilt shopping areas, to make shopping safer against the growing menace of the motor vehicle.

The shopping centre at Stevenage was one of the first modern shopping centres to be designed to exclude the car completely from its shopping pavement, and since then there have been many more.

A suitable size for a group of shops larger than the parade discussed earlier

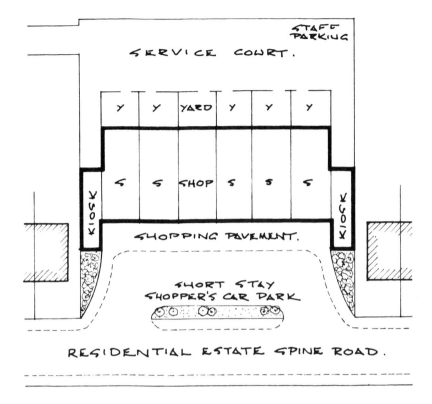

Figure 10.1   Shopping parade

is one of between 12 and 20 shops. This would be a suitable size to service a centre for a small neighbourhood or to provide the necessary shopping facilities for a mainly residential development of 5000 to 6000 people. It could also form part of an urban renewal scheme in an established town.

Basic Requirements for the Shopping Precinct   The following are the basic requirements and general principles to be considered when planning a shopping precinct.

(1) The site should be the focal point in a residential area. It should be near the centre provided that it is directly accessible from the main road network.
(2) Vehicular traffic should be excluded from the pedestrian area.
(3) Shop fronts should face on to a central pedestrian concourse with shopping pavements leading on to it. These should be wide enough to permit free movement but not wide enough to isolate the shops on one side from those on the other side.
(4) Shops should be serviced from the rear (above or below), leaving the front exclusively for salesman–customer contact.

91

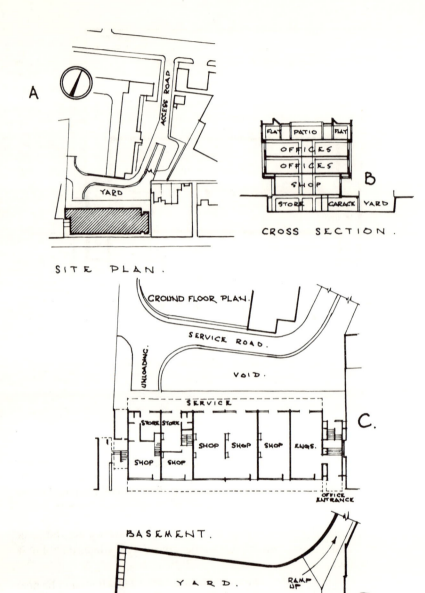

A

ACCESS ROAD

YARD

SITE PLAN.

FLAT PATIO FLAT

OFFICES

OFFICES

SHOP

STORE GARAGE YARD

B

CROSS SECTION.

GROUND FLOOR PLAN.

SERVICE ROAD.

UNLOADING.

VOID.

SERVICE

STORE STORE

SHOP SHOP SHOP RLNGS.

SHOP SHOP

C.

OFFICE ENTRANCE

BASEMENT.

YARD.

RAMP UP

GARAGES

D.

SHOP STORES

Figure 10.2 Shopping parade

92

(5) Service roads should, where possible, be 'one way' and should be sufficiently wide to allow one vehicle to pass another standing outside shop premises for unloading.

(6) If a service area has to be a cul-de-sac, adequate turning facilities must be provided.

(7) Parking for shoppers should be adequate and clearly visible from approach roads. Separate provision should be made for the cars of people working in the precinct. (Servicing from above or below is dealt with in chapter 10.)

(8) The footway from the car parks to the concourse and the shopping pavements should be free from interference by vehicular traffic and ramps should be provided at changes of level.

(9) There should be direct pedestrian access from bus stops and any other transport stops or terminals.

(10) The shops should be compactly grouped in terraces without breaks in the shopping frontages.

(11) Where shopping frontages change direction the shops on each frontage should be clearly visible from the other.

(12) Ornamental features and decorative furnishings should be used sparingly. They should be vandal proof and sited so that they can be seen from as many shops as possible.

(13) A limited number of benches should be provided for shoppers to take a short rest but they should not be encouraged to sit for long periods.

(14) Potted plants and shrubs should only be considered if provision can be made for them to be tended and replaced.

(15) Where there are alternative routes within the precinct these should be generously sign-posted.

(16) Public toilets should be discreetly sited but at the same time convenient to the precinct.

Planning a Precinct An economic survey and local investigation will reveal the likely catchment areas and the number of potential customers living in the area. It will also provide information on types of employment, wage levels and other relevant material to enable an assessment to be made of the spending power in the neighbourhood. It was suggested earlier that owing to the obvious changes that are taking place in the pattern of retail trading, a conservative approach is essential to leave some margin for a downturn in local trading. At the same time the local population should not be deprived of shopping facilities and the developer should not miss the opportunity to carry out a profitable development because of excessive caution.

Individual shops These need not be of standard size, although some uniformity is desirable. If a supermarket is prepared to take premises in a new development it is often worth while designing a building to their specification, provided it fits into an overall plan and does not dominate the precinct to the detriment of other occupiers.

Shops facing on to a central concourse should be of a standard size, about 6 m × 18 m, and if there is either basement or upper-floor storage the depth can often be reduced. They should be constructed so that they can be interconnecting, with two or more shops occupied by one retailer. In cases where there are narrow branches off the main concourse leading to car parks, bus stops and the like, these can form pedestrian pavements and be lined with small shops, which may have frontages down to as little as 3.6 m.

Notwithstanding the basic principle that there should be vehicular access to the rear of all shops, it might be found practicable to service this type of small shop from a service footway, with goods brought by trolley from the service road. This may be a solution on a restricted site in an urban renewal scheme.

Paved areas The main shopping concourse should be wide enough to encourage shoppers to stay and look at all the displays on one side before crossing over, but not so wide as to suggest remoteness; 9 m is considered suitable. The shopping pavements can be reduced to about 5.4 m and still permit the free flow of pedestrians in both directions.

Servicing One-way traffic will usually reduce the land required for service roads and will allow traffic to flow more freely. The minimum width should be 6 m but on a restricted site it can be slightly less.

Parking The amount of parking provided will depend on many factors, which will include the following.

(1) In a mainly residential area many of the shoppers will come on foot, and in this case an allocation of between 1.5 and 2 spaces per shop is likely to be sufficient.
(2) For a shopping precinct serving a predominantly country area where most shoppers come by car the allocation should be higher.
(3) In an urban renewal scheme the parking may be centralised and any parking provided would make its contribution to meet the overall provision. Forecast figures for parking in 1980, put forward by the Multiple Shops Federation,[1] are between 4 and 6.5 cars per 100 m^2 of retail area, according to the level of car ownership in the region.
(4) In the case of a shopping development which includes living accommodation or flats above, separate provision will have to be made for garaging and also for visitors' parking. This should be kept separate and be strictly allocated. People working in the precinct should not be allowed to use the shoppers' car park unless it is deliberately made larger for the purpose. Even then parking should be by permit only or controlled by some other means.

The entrance The main entrance should be off an existing street, preferably near the peak shopping position. It should have an appearance of some importance and should attract the casual shopper's eye without resembling the entrance to an amusement arcade. There should be something interesting to see immediately inside the entrance and a display inside the precinct which can be seen from the entrance. Most of these and other ways of attracting customers into the precinct will be the province of experts in marketing but the developer is not free from all responsibility in this field. He should meet any additional structural requirements and pay particular attention to the siting of points of access.

Residential accommodation Reference has already been made to points for and against providing living accommodation as part of a shop development.

Independent flats over a shopping precinct may be a very appropriate form of development provided the access to the flats does not break up the shopping frontage. It must be well situated, sufficiently impressive for the rent level being asked and adequate for the number of people likely to use it. The usually untidy area at the back of the shops should not be visible from the flat approach. The front of the flats should not be flush with the front of the shops but should be set back to allow the front part of the shop to be decked over to form an open terrace or balcony in front of the flats. If the flats are set back 6 m, the flats facing each other across the concourse will be at an acceptable distance from each other from the point of view of being overlooked.

Examples of precincts are given in figures 10.3 and 10.4, and a compromise showing shops on three sides of a quadrangle with a car park in the centre is shown in figure 10.5.

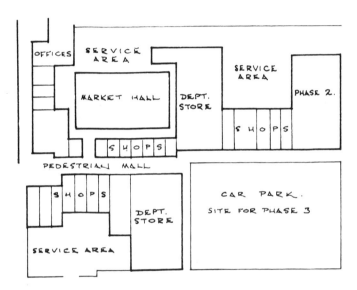

Figure 10.3   District shopping centre

Most towns now have some pedestrian-only shopping. Some have been able to exclude vehicular traffic from the traditional shopping streets. This method seems to have been successful where the usual difficulties of making alternative arrangements for servicing the shops have been overcome.

One or more shopping precincts have generally been built in most towns as part of a scheme of urban renewal. Some have been successful, some have almost totally failed, but most are surviving, although in many cases the shops are not commanding the rents the developers had hoped for.

Out-of-town Shopping and the Hypermarket

This comparatively new form of shopping was introduced into the United

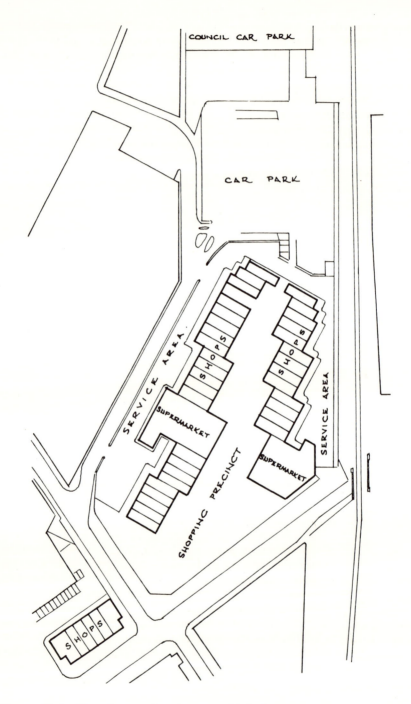

Figure 10.4   Shopping precinct in an established town

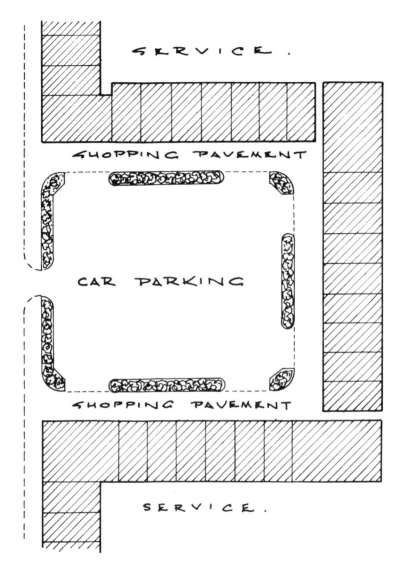

Figure 10.5 Shopping quadrangle with central parking

Kingdom from the United States in the 1960s. One of the first, developed by the firm of GEM, was built in West Bridgford on the outskirts of Nottingham. Woolworths then formed its subsidiary company, Woolco, to develop this form of shopping, building the first Woolco store at Ordby, near Leicester. There are now a number of Woolco stores, including one at Bournemouth and one at Thornaby. In the past few years the French firm of Carrefour has broken into the market and introduced the word 'hypermarket' to describe a self-service store with a sales area of not less than 2300 m². Since the original conception, sizes here increased to 4600 m². For some years it has been the practice for the Secretary of State for the Environment to 'call in', for his own determination, planning applications for 'large' stores—at present (1980) this applies to stores over 9200 m².

The hypermarket is a free-standing building, purpose built to provide a very large unbroken sales area. A wide range of food and other merchandise is sold, usually at a discount. It is located out of town, preferably in open country, and there are generous car parking facilities. There may also be additional services in the form of a petrol filling station and a rapid tyre-changing service and possibly a garden centre. Some also have such additional services as hairdressing and shoe repairs.

The hypermarket will usually provide a terminal for the local bus service but at the same time it depends on the motoring public for most of its trade and, to a certain extent, on the two-car family. In estimating the number of potential customers the first assumption is that those living nearest are the most likely to use the store. For this reason the catchment area is divided into zones, each zone being a different distance away from the centre. The 'distance' is measured in minutes of travel time. The zones are usually at 5 minute intervals of driving time and are from 10 minutes to 30 minutes. When measuring these distances it is important to use a family car and to do it during off-peak periods. Since this form of retailing is usually superimposed on an existing pattern of shopping facilities, the promoter has to break into that pattern—the process is called 'market penetration'. Evidence so far indicates that a hypermarket offering food at competitive prices may attain up to 15 per cent of the total food purchases by car owners within 10 minutes' drive and up to $7\frac{1}{2}$ per cent of those between 10 and 20 minutes' drive time away. It is generally accepted that a hypermarket needs a catchment area with a population of 100 000 within 30 minutes' drive but that a large proportion of their takings must come from those living within 15 to 20 minutes' drive. A typical site layout is shown in figure 10.6.

An economic survey of the first year's trading of a south country hypermarket produces some helpful figures. The hypermarket has a total floor space of 102 000 ft², of which 50 000 ft² is sales space (about 4550 m²). In 1974/5 takings were estimated at £10 million, which represents £2200 per m² of shopping space per annum, and £21.50 per head of population in the 20 minutes' drive radius. (This figure is slightly distorted because some customers actually come from further afield.) At the end of the first year the store was attracting 26 000 groups of shoppers per week, of which 95 per cent came by car. These were spending an average of £10.60 per visit, whereas the remaining 5 per cent only averaged £6.50 per visit. Of the customers coming by car those who lived within 5 minutes' driving distance spent an average of £6.30, and those travelling 30 minutes or more spent an average of £14.50.

With regard to the provision of parking facilities, the Urban Land Institute, on the basis of experience in the United States, recommends 5.5 spaces for each 1000 ft² of floor space which, when converted, is one car space to every 16 m² of shopping floor space. Continental experience indicates that this is

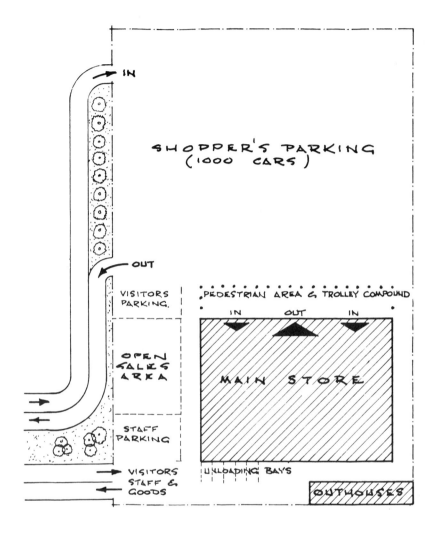

Figure 10.6 Hypermarket

low and that one car for every 9 m² is more appropriate. Staff parking needs will depend on the location of the store in relation to public transport facilities.

Cash-and-carry and Discount Stores

The cash-and-carry type of store began as a wholesale warehouse where small retailers, café owners and the like could purchase comparatively small amounts of goods for resale. The sales technique was to offer goods direct from bulk packaging with little or no display. The goods were paid for on the spot and carried away by the purchasers. Since they were warehouses for

99

wholesale distribution they normally fitted into light industrial estates but they have now become much more like supermarkets and attract as much activity as a retail store. This has resulted in some planning authorities refusing applications for such use.

Another product of the times is the discount store. These are often established in a hangar on a disused airfield, where furniture, carpets, electrical goods and the like are sold for cash at large discounts. They have grown as a result of the recent economic climate and may not survive in the economic climate of the next few years. If they are here to stay they too will make their contribution to the distortion of the pattern of retail trading and there are indications that this will be at the expense of the more conventional type of trader.

REFERENCE

1. — *Car Parking and the Shopper* (Multiple Shops Federation, London)

11 TOWN CENTRE DEVELOPMENT AND RENEWAL

This is the most difficult chapter to write at the present time owing to the uncertainties associated with town centre development. In the 1960s there was considerable activity in this field, but because of the enormous increase in costs, comprehensive town centre renewal may have become too costly to contemplate except on a piecemeal basis. The situation has been further aggravated by the changes which are taking place in people's shopping habits, and the growth of out-of-town shopping, which is increasing in momentum. If this continues and gathers more strength there may have to be a complete 're-think' on the whole concept of the town centre and its function as a principal shopping centre. The situation throughout the United Kingdom today is that the town centres of a number of the latest new towns are in the course of construction and some central area schemes are still in progress, but most if not all of these were first planned 10 or more years ago. Some others have been postponed indefinitely and some have been virtually abandoned, notwithstanding the considerable amount of money that has already been spent. The attitude of the developers has been to cut their losses and withdraw from a situation where they see little prospect of a scheme ever reaching a successful conclusion.

Building a town centre for a new town is a very different matter from rebuilding the centre of an existing town and must therefore be considered separately. The third type of centre worthy of separate consideration is an existing town which has to be enlarged and improved to cater for a planned increase in population.

THE TOWN CENTRE OF A NEW TOWN

The first batch of new towns built in the 1950s is now generally classified as mark I and it was, in the main, designed on the neighbourhood principle. The towns have a town centre surrounded by a cluster of neighbourhoods, each with its own neighbourhood centre and some also with subcentres. In the early 1960s there was a definite move away from the neighbourhood concept. The outcome of this move is to be seen in the design of Cumbernauld new town, Dumbartonshire, which was based on the principle of centrality. The requirements for this town centre, given in the preliminary planning proposals published in 1958, were

'The central area of the town will provide sites for shops, offices, public, cultural and recreational buildings and it is considered that there should also be some housing in high blocks within the area to ensure that it maintains a lively atmosphere even when the shops and offices are closed.'

To meet these requirements the centre was designed with the object of locating within one single multi-decked structure all the social, governmental, cultural, educational, commercial and shopping functions of the town. The centre is served by pedestrian ways and serviced by car parks, roads and delivery bays situated at lower levels and reached by escalators, lifts, stairways and ramps.

This compact form of development, shown in figures 11.1a and b, was meant to provide greater flexibility than was possible in the mark I new towns but there is still some doubt whether this flexibility has been attained. It was necessary at the outset to persuade private enterprise and local and central government to accept a radical change in attitude towards types of tenure, space allocation and design. Their full co-operation in accepting the complicated structural discipline required was essential to the smooth operation of the centre.

The design of Cumbernauld was new when the town was first built and even in the 1970s there were still doubts about its success. These doubts are well expressed in the following quotation from a study of the town.[1]

'The Town Centre can be appreciated only as a sequence of visual impressions, as it looks so different from various viewpoints. It has a striking many-sidedness like many of Le Corbusier's buildings, and, as in Le Corbusier's buildings, the inner and outer spaces penetrate each other inextricably. The body of the Town Centre has been hollowed out in every direction, hence its draughtiness. Almost all the ornamentation of the Town Centre is provided by the paint spraying on walls and doors by young hooligans.'

THE ENLARGEMENT OF THE CENTRE OF AN EXISTING TOWN

The co-operation between local authorities and major authorities such as the Greater London Council operating under the terms of the Town Development Act, 1952, has resulted in the expansion of many provincial towns. In most cases the results have been good. It has revitalised what in some cases had become a slightly run-down community by introducing new industry and population. This has resulted in the need to rebuild or enlarge the town centre. Proposals of this kind are often the cause of objections by the local people opposed to change.

The expansion of Andover, Hampshire, is a typical example of co-operation between the authorities. In 1961 the population was 17 000 and the plan provided for expansion to 48 000 in the 20 years to 1981. It was anticipated that this increase in population would demand increased shopping facilities and to provide for this increase about one-fifth of the traditional centre of the town was redeveloped on the lines shown in figure 11.2. Although there are two large supermarket units their bulk does not dominate the centre owing to the presence of a six-storey office block at the focal point of the development.

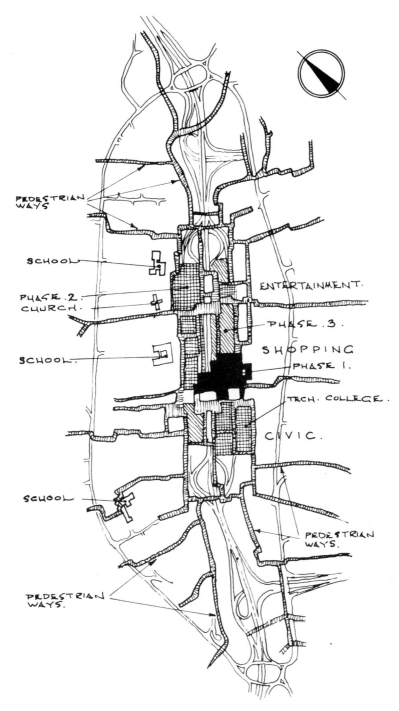

Figure 11.1(a)   Cumbernauld, plan and section

103

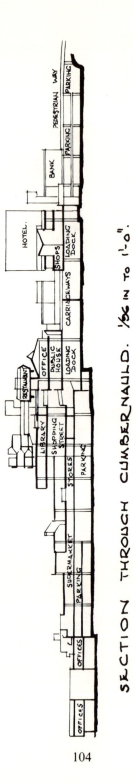

SECTION THROUGH CUMBERNAULD. $\frac{1}{32}$ IN TO 1'-0".

Figure 11.1(b)   Cumbernauld, plan and section

104

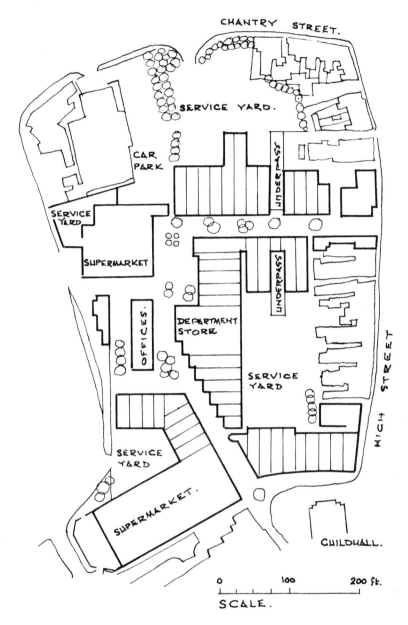

Figure 11.2   Andover shopping centre

The relationship between the new development and the other central area uses is shown in figure 11.3. Andover is a pleasant market town with a funnel-shaped High Street and market overlooked by a Greek-revival guildhall, and

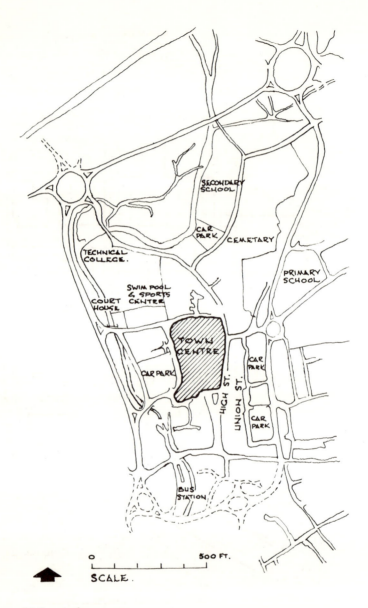

Figure 11.3   Andover

the new development does not detract unduly from the eighteenth century character of the town. Another interesting example of a smaller scheme is given in figure 11.4.

Many town expansion schemes are now being run down and the financial resources intended for them are being diverted to inner city redevelopment schemes.

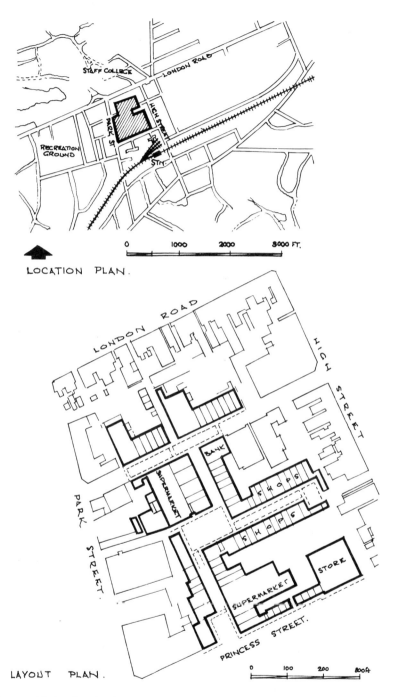

LOCATION PLAN.

LAYOUT PLAN.

Figure 11.4   Camberley

107

TOWN CENTRE RENEWAL

This has been the subject of two government planning bulletins both issued in the early 1960s.[2,3] Between them they outline Government policy towards town centre renewal and give official approval, indeed encouragement, to the co-operation between local authorities and private developers in planning joint redevelopment schemes. They do emphasise, however, that the local authority should retain the initiative. The following extract from the foreword of the first of the two publications sums up the attitude of the Government at that time.

'The town centre plays a vital role in the life of a town. It should be attractive, convenient and efficient. But today many town centres are threatened with obsolescence, decay and congestion. This is a threat not only to the future prosperity of the town centre itself but to the whole community of which it is a focal point. The renewal of the centre may well become one of the most burning and also the most important of local issues . . .'

The bulletin goes on to suggest how the difficulty of land assembly can be overcome by taking the following steps.

(1) The local authority defines a (then) CDA and prepares a scheme in agreement with a development company.
(2) The company acquires as much land as it can by agreement, the local authority using powers of compulsory purchase only as a last resort.
(3) The company would then transfer the freehold of its lands to the local authority, who would grant a long lease in return.
(4) The terms of the lease are to provide for periodic review of the ground rent.

It goes on to state that the Minister is not opposed to the use of compulsory powers for this purpose but emphasises that a scheme must meet all the statutory requirements, be sound and in the public interest, and that it should be judged on its planning merits.

The objectives of a scheme can be examined under the following four main headings.

Function

This can best be established by seeking the answer to a number of questions.

(1) What are the future size and purpose of the centre to be?
(2) Is there need for expansion?
(3) Ought it to be a regional centre?
(4) Does it have or will it attract other functions, such as national exhibitions centre, government offices?
(5) What is an appropriate proportion of residential use in the centre?

Layout

The questions with regard to layout should include

(1) How are the main uses distributed at present?
(2) Are the shops conveniently placed?
(3) Are there intrusive uses in the shopping parade?
(4) Are the bus stops conveniently sited?
(5) Are there enough car parks and are they easily accessible?
(6) Does the architecture and siting of the public buildings reflect their importance?
(7) Is there scope for reshaping the present layout?

Circulation

Traffic movement and circulation are interrelated with the function of layout but they should not overrule all other considerations. A satisfactory traffic system is fundamental to the success of a town centre renewal. The primary objective must be to recognise conflicting types of movement and traffic and then sort them to provide adequately for each type, and in particular to separate pedestrians and vehicles so that both can move freely and with safety.

Character

Every town has a character of its own; this is often an elusive quality but it is what makes the town. Underlying all other objectives there should be the desire to retain and enhance individuality. The slow, piecemeal development which gave character to the centres of many towns is a thing of the past but this does not mean that the way to redevelop a town centre is to demolish everything, sweep clean and start afresh.

The second bulletin concerned itself with the financial aspects of schemes, the importance of comprehensive economic surveys and strict control over costs and in the apportionment between public and private uses.

An economic survey should have the following four aims

(1) to assess the town centre in commercial terms
(2) to establish its status as a shopping centre
(3) to note its present character
(4) to judge its prospects for profitable expansion and redevelopment.

It should cover both existing conditions and forecast likely trends.

The scope of the survey will depend on local factors but should take the following lines.

Town Characteristics

Population growth and trends
Nature, place and type of work of majority
Average earnings and spending power
General level of prosperity

Shopping-centre Characteristics

Status in the region
Other competing regional centres, both existing and proposed
Present drawing power and effect of competitive redevelopment elsewhere
Innate local advantages under subheadings of amenity, character, conveniences and tradition
Physical condition and age
Buildings of architectural or historic interest
Number, size and use of buildings making up the shopping centre
Total floor space devoted to retail outlets

Demand

Nature of present ownership
Possibility of increase or decrease in space
Conflict between size and position
Needs and deficiencies
Likely movement of traders affected by the proposed scheme
Likely increase in turnover
Use of upper floors
Demands for office space

Location Requirements

Siting of focal points
Reservations for 'magnet' traders in key positions
Siting of ancillary users, including hotels, entertainment centres and services

Implications

The effect of redevelopment on other commercial interest in the town
The effect of competition

Estimates

This will ultimately be a major valuation exercise but it may be useful to have some preliminary figures, including the approximate cost of

Acquisition
Site clearance
Construction work
New rental values
Ground rent

The following additional information may be of value and should be appropriately recorded as part of the survey

Market trends
Sources of private capital
Location and needs of particular traders

The outcome of this economic survey should enable a local authority to decide on the scope and size of any scheme. The basic costs to be met by the local authority will vary with each scheme but these will in most cases include

Land assembly (may be part only)
Clearance and preliminary development
Provision of roads, services and other public improvements

There may also be civic buildings for the use of the authority but these should not be at the expense of the commercial parts of the scheme. The cost of the commercial development will normally be met by private enterprise. Although the public and private sectors will be completely separate as accounting units, they should not be opposing forces but should interact to enhance land values to their joint benefit.

Profit should not be the main consideration but economics must be taken into account. This does not mean that major planning objectives should be abandoned to achieve a better financial return, but one of the aims of a redevelopment scheme must be to open up opportunities for profitable development.

In the light of the importance of the financial aspects of a scheme, both the profitable and the unprofitable parts should be identified at an early stage and this can be done by an economic analysis on the following lines.

The total area is divided into parts, according to their proposed use in the new scheme. These will fall into two main groups

Land available for commercial development
Land required for public use

There may be some justification for having a third classification to cover all the land required for roads, open spaces and even car parks. Costs in this analysis are restricted to the site and will *not* include building costs. They should be based on values appropriate to acquisition by compulsory purchase and should include an estimate of disturbance and other heads of claim. The cost of demolition, site clearance and preliminary development will also be included.

A number of schemes submitted to the Minister in the early 1960s were analysed on this basis and some of the factors causing losses were found to be

(1) areas included existing buildings of high value not ripe for development
(2) areas included little underdeveloped or underused land

(3) a large proportion of land was appropriated for public use at comparatively low values
(4) failure to provide opportunities for commercial development
(5) inefficient site planning in failing to make the optimum use of the best commercial sites
(6) extravagant layout and the heavy cost of preliminary development resulted in a reduction in the area available for remunerative use
(7) site conditions were difficult and the cost of distributing underground services was high.

Another government publication that has contributed substantially to thinking on town centre renewal is the Buchanan Report[4] which deals with the problems of traffic and congestion. The traditional centre of a town is often situated at a main crossroads with the main shops and other central area activities firmly established along the roads leading to the crossroads. The market is also likely to be close by. This leads to congestion at the centre if the traffic is not properly controlled. Inefficiency in traffic control in a town centre can usually be attributed to one or more of the following causes

(1) traffic passing through the town centre because there is no alternative route
(2) inefficient layout of streets for traffic purposes and multiplicity of street intersections
(3) confusion of building uses, creating and attracting both pedestrian and vehicular traffic, causing the roads to be used for more than one purpose
(4) inadequate provision for unloading, waiting and parking of vehicles
(5) the virtual attainment of saturation point in the capacity of some urban roads to take more vehicles.

To remedy the conditions which arise from these causes, the report sees a need for the following.

(1) The progressive remodelling of the town centre, and of the areas surrounding it, into a series of environmental areas. In a small town, the central area itself might be entirely confined to one of these, but in larger towns it would occupy several environmental areas, which would perhaps tend to be differentiated from each other in terms of their predominant land uses.
(2) Diversion from the central area of all traffic which has neither origin nor destination within it.
(3) Definition of roads in classes (by function) according to the nature of the traffic which they may be expected to carry; this will usually depend upon whether they carry only traffic generated by the buildings along their own length, or whether they carry traffic generated over a wider area, which could in some cases include the entire urban region.
(4) Free flow of traffic, entailing elimination of all unnecessary intersections, provision for waiting vehicles, and avoidance of conflicting traffic uses.
(5) Convenient circulation for pedestrians, particularly in shopping areas.
(6) Adequate and well-distributed system of car parks.
(7) Retention of existing streets where these can be suitably incorporated

within an improved street layout or possibly converted for use by pedestrians only, and execution of proposed changes at reasonable cost and by orderly stages.

Town centre renewal may be planned as one comprehensive scheme but it can often be broken down into a number of smaller schemes, each contributing its part to an overall plan. This can be successful provided there is a master plan and that direction and co-ordination from the top is of the highest order.

The remainder of this chapter is devoted to a review of a large new shopping centre, forming part of an enlarged town centre, and making a substantial contribution to the shopping facilities provided.

THE ARNDALE CENTRE

The town centre redevelopment industry, particularly in the field of town centre renewal, is almost completely inactive. Few if any new schemes are being started and many have been postponed indefinitely or completely abandoned. This situation is not generally due to a lack of appreciation by the shopping public but to the enormous increase in the costs associated with this kind of development. A covered shopping centre is in many ways an asset to the town and it will be a great pity if this type of development has priced itself out of the market for good.

Arndale is one of the best-known names in this modern form of town centre development. There are about 20 centres in different parts of the country and one of the largest has just reached completion at Luton, Bedfordshire.

Report

The Developer The development, which is known as the Luton Arndale Centre, was carried out by a partnership between the local authority (then Luton Borough Council) and Town and City Properties Ltd. The site was assembled by the joint efforts of both parties with the local authority using powers of compulsory purchase in many cases. In all some 600 business interests were purchased. The whole scheme will have taken 13 years and the chronological table, reproduced as appendix F, is worthy of study. Since May 1973 the development has progressed in stages along the line of the central mall, and as each stage has been finished it has been opened to the public. The whole scheme is now complete.

The local authority now holds the freehold but the site is on a long lease to Arndale (Town and City Properties) which as landlord now carries out the day-to-day management. (Arndale was the name of the company that originated this type of town centre. They have been taken over by Town and City Properties Ltd, but have wisely retained the name 'Arndale'.) The market hall is leased to the local authority, which handles the letting of the stalls. The rooftop parking has also been taken over and it forms part of the system of car parks in the town and is not therefore exclusive to shoppers at the Arndale Centre.

The Site   The development stands on about one acre. It is almost an island site, as can be seen in figure 11.5. It is shown as fronting on to George Street, which may be strictly true but is not structurally so. A number of the shops and larger stores have retained their frontage and much of their original selling area and have extended at the back into the new development, having display frontages and accesses on the shopping pavements within the centre.

The Locality   The shopping centre forms part of a town centre complex and both the railway station and the bus terminal are within short walking distance of the site.

Luton has a population of 160 000; it is almost entirely dominated by the two large motor manufacturing plants of Talbot and Vauxhall and is therefore very much dependent on the fortunes of the car industry. It is situated in the most rapidly growing area in the United Kingdom and is surrounded by new or expanding towns, including Milton Keynes, which is being built about 32 km (20 miles) to the north-east.

In an area with so many modern town centres it is almost impossible to establish a catchment area, but the landlords conducted a number of counts some time ago and they estimated that at that time there were 400 000 people in the centre during an ordinary week. Recent checks indicate that the figure has probably reached half a million.

Access and Circulation   The shopping pavements are covered and fully pedestrianised and consist of a central mall widening into a square at each junction and change of direction. There are three branches off the central mall, each leading to a pedestrian access off the George Street frontage. There is further access at the western end of the central mall off an open pavement running between a hotel forecourt and the public open space at the corner of Bridge Street and Manchester Street. Access is also possible and indeed encouraged through a number of stores with frontages to both the shopping pavements within the centre and the traditional shopping street outside. The mall and other shopping pavements are 10.9 m wide and circulation is to some extent controlled by the careful positioning of statuay, fountains, plant vases and seating.

Space in the centre of the shopping pavements is let for the erection of display stands when exhibitions are held. These help to attract more people to the centre and if the stands are well sited they encourage more circulation by the public, to the benefit of both the exhibitors and the permanent shops. In order that the exhibition stands should not dominate the scene at the expense of the permanent shops, a height limit of 2 m is imposed whenever practicable.

There is a gallery around parts of the central area with smaller types of shops, occupied in the main by business not dependent on passing trade. A number of the occupiers of the bigger shops below have extended into gallery shops, and this has increased the number of people who now use the gallery, and the circulation generally has improved at this level.

Figure 11.5 Arndale Centre, Luton

115

The servicing of the shops within the centre is from a security-controlled access at the end of Melson Road, to the north-west of the market hall into a large semibasement service area under the main buildings. It has headroom of 4.4 m. The roads in this area are directly below the main pedestrian routes on the shopping floor above. They are 7.2 m wide. Off these roads are loading bays serving store rooms occupied jointly with the premises above. A number of the occupiers of the bigger shops and stores have their own internal goods lift and others use the goods lift provided for general use. The stallholders in the market hall are serviced in the same way as the rest of the retailers but their storage space is allocated according to the type of trade and the amount of space required. Traffic circulation in the service area is one way, with the exit on to Library Road and under the forecourt to the hotel.

Other Uses In addition to the shops and stores there is a covered market with 145 stalls which has access from the street and off the central mall. There are two office blocks, one of 4600 m^2 and the other 1800 m^2. There is a new luxury hotel with 150 rooms, operated by Thistle Hotels. This has a forecourt which has direct access to shopping pavements at the same level. There is a pub with access and bars at both ground level and off the central mall. This end of the mall is fully glazed to give a view of the church and the surrounding gardens, and it is obviously a popular place for people to sit. There is a petrol filling station in the complex.

Garages and Parking There is some parking for operational vehicles alongside the loading bays associated with the basement storage, but this is not actively encouraged since it tends to hamper the free flow of traffic, opposing the main function of the service area. There are two multistorey public car parks with a total capacity for 1500 cars. These are linked to rooftop parking over the centre and this is served by lifts from the shopping level. It provides for an additional 720 cars.

Services All the usual services are available. The pipes and cables for these services are under the shopping pavements and are threaded through the lattice formation of the prestressed concrete beams which support the floor. In this way they are directly accessible from the service area. The whole centre is air-conditioned. A 24-hour security guard operates in the service area and there are security patrols on the selling floors. There is a nonstop cleansing service and refuse is collected from the service area.

Landscaping Externally there are plans to plant and landscape the open spaces at the western corner of the development but there are very limited opportunities elsewhere in the scheme. There is a small garden in the triangle outside Woolworths and a few shrubs and vases outside the market hall.

Some variation in the paving has been attained by using granite setts to form a simple mosaic pattern in the paving stones but this has little impact. Inside, the paving of the mall and other pedestrian routes is more ambitious. The colour, texture and size of the tiles create quite a pleasing effect.

There are a number of vases containing potted plants and shrubs and these are generally well sited in groups. There are two fountains and an impressive piece of sculpture, representing a group of flamingoes—the whole piece stands about 4 m high.

The two outstanding features of the shopping areas are the excellent lighting, which has been planned with considerable taste, and the cleanliness of the floors with the almost complete absence of litter.

Values   It was estimated in 1974 that the capital cost of the whole project would be £25 million but the final figure is likely to be higher. The local authority's investment in the scheme is approximately £10 million. Rents are high, as is to be expected in such a centre, but the indications are that the returns make it worth while. The last phase is now almost complete and most if not all of the shops have already been let. The developers are reluctant to give details of rent because they are the subject of negotiation. This is quite understandable and the little information obtained would tend to confuse the issue.

Conclusions

This development was started when it was generally believed that there was a fortune to be made out of this kind of development. Some successful centres have been built and there have been some failures, but the Luton Arndale Centre seems to have passed through a difficult period of depressed values and a large scheme of development has been completed which not only conforms to the original plan but which, it is understood, is on time. The indications are that, provided there is not a drastic fall in earnings in the motor manufacturing industry, it will live up to the claims made for it.

The air-conditioning and other first-class facilities provided in an Arndale Centre take most of the hard work and discomfort out of shopping but we shall only retain this form of shopping if people can *afford* to shop in these highly pleasant settings. Notwithstanding the expensive surroundings, prices must be competitive or in the long term people will look elsewhere, particularly if money becomes short. This is an ambitious scheme which looks like being successful.

REFERENCES

1. Ferdynand Zweig, *The Cumbernauld Study* (Urban Research Bureau, London, 1970)
2. Department of the Environment, *Planning Bulletin No. 1, Town Centres 1962* (HMSO, London, 1962)
3. Department of the Environment, *Planning Bulletin No. 3, Town Centres 1963* (HMSO, London, 1963)
4. Colin Buchanan, *Traffic in Towns* (HMSO, London, 1963)

12 INDUSTRIAL DEVELOPMENT

Most of the development agencies considered earlier in this book take some part in industrial development, as the following list illustrates.

(1) *Central government* Although much reduced in recent years, there are still a number of ordnance factories, dockyards and research establishments that are occupied by government departments. These departments are usually advised by the Public Services Agency (PSA).

(2) *Nationalised industries* Most of these have their own estate department responsible for providing accommodation. One exception is the GPO which, although no longer a government department, is still advised by the PSA.

(3) *New town corporations* One of the most important components of a new town is its industrial estate. This kind of estate is discussed later in this chapter.

(4) *Industrial development corporations* Although limited in number these are not unlike new town corporations, except that their primary function is to establish an industrial estate. They are not concerned with other forms of development, except to the extent of providing ancillary services. The Teesside industrial estate considered in some detail later in this chapter is an example of this form of development.

(5) *Local authorities* The development of industrial estates is very much the concern of local authorities if they are involved in a scheme of expansion under the Town Development Act, 1952. In these circumstances their requirements are similar to those of new towns.

(6) *Private developers* Industrial estates developed by private enterprise contribute substantially towards meeting the demand for small factories, warehouses or workshops for sale or rent. The developer can usually offer greater flexibility both in the size of the unit and in catering for the requirements of specialised industries.

In the context of this book industrial development is seen as the development of industrial estates; outside its scope are large industrial complexes such as those operated by the National Coal Board, gas and electricity boards and large motor manufacturing plants.

INDUSTRIAL LOCATION

The control over the location of industry at local level is now an important part of town planning legislation, taking its present form in 1948. The policy adopted by the Government was largely influenced by three different publications. The first of these is the Barlow Report,[1] the second is the Scott Report[2] and the third is a paper published in 1944 by Sir Patrick Abercrombie called *Planning and Reconstruction*; each makes a contribution towards the planning for industrial location. In the mid-1940s many industries were still operating where they were first established over 200 years earlier. This is true of such basic industries as wool, cotton, shipbuilding and steel and there were also what the Barlow Report calls 'link industries'—industries which supply parts for other industries and therefore need to remain close to them.

There are arguments both for and against the policy, but it is still the aim of the Government to provide more jobs and greater diversification of employment by encouraging industrialists to take new factories to areas of high unemployment, particularly where some of the old basic industries are running down. In certain areas there are tax concessions and other financial inducements to attract industry and in extreme cases a development corporation has been set up by statute and charged with the duty of establishing an industrial estate.

The other main competitors in the market for new industries are new town corporations and local authorities expanding under the Town Development Act, 1952. In addition to trying to attract new or expanding industries they are both trying to attract nonconforming industries from their present sites—established industrial users in nonindustrial areas who have the right to stay but are restricted in such matters as change of use and limited extension of buildings.

The importance of locating certain types of industry in traditional areas has become less marked with the increased mobility of both goods and labour and, in the case of fuel, by the extensive use of electricity. However, where there is a pool of experienced labour in the locality certain areas still attract specialised industries.

With automation and the high level of mechanisation in industry today, a large proportion of industrial processes are carried out by semi-skilled labour, and skilled craftsmen are only needed to maintain the machine tools and to set them up for the semi-skilled to operate. These conditions make most industries much less tied to a particular area than in the past. If an industrialist plans to move to a new area it is often the case that he will only take with him the nucleus of a work-force consisting of management personnel and technical staff, and will recruit most of his main labour force locally even if training will be needed. The Government is now trying to encourage the revitalisation of inner city areas, and this will mean establishing new industries and providing facilities for traditional industries to stay if they are suitable for inner city areas. This about-turn in government policy is likely to have far-reaching effects, and the particular problems of inner city development are dealt with later in the chapter.

CLASSIFICATION OF INDUSTRY

Industrial buildings are required to house a wide range of plant and machinery to support many different industrial processes. The materials used can be small or bulky, light or heavy, expensive or comparatively cheap. Some buildings provide little more than protection from the weather and hardly need to be locked up. Others need to be warm and well insulated and in some industries may need air-conditioning installed—these buildings need to be fully secure against possible break-in.

To exercise effective control over land use for such a wide range of industrial activities, the Use Classes Order referred to earlier and summarised in appendix B defines industrial buildings and also classifies them as

> Class III Light industrial buildings
> Class IV General industrial buildings
> Classes V to IX Special industrial buildings

The normal industrial estate is likely to have light and general but seldom special industries. The planning authority will indicate the type of industry it is prepared to approve in a given location.

Although it is becoming more common to develop separate estates for warehousing it is not unusual for a planning authority to approve an application for warehousing on a light industrial estate. There may also be wholesale distribution depots and cash-and-carry stores.

ALLOCATION OF LAND FOR AN INDUSTRIAL ESTATE

In planning for industry, particularly in a new town, the target population should be taken into account since this must have some bearing on the number of people likely to work on the industrial estate. The ideal from the planning point of view would be for all the working population in a town to be employed in the town. This will never be so but it can often be a reasonable assumption that the number of people who come into the town to work compensate for those going out. It has been estimated that about 20 per cent of the population of a new town work in industry. This would mean that in a target population of 60 000 there would be 12 000 who would seek employment on the industrial estates.

A survey of Trafford Park industrial estate conducted in the early 1940s showed a figure of 45 workers per industrial acre and on the basis of this figure the Government recommended 50 to 60 as being appropriate for new developments. To provide sufficient accommodation for the 12 000 workers referred to above, between 68 and 80 hectares are required. The distribution of this amount of land and its location depend on many local factors and site characteristics but there would normally be two large estates each of 25 hectares or more with the remainder devoted to light industry in small sites conveniently situated near the residential areas and even smaller sites for service industries in the residential areas.

Site Coverage

In the first instance new towns adopted low densities and in the early 1960s there were only about 75 workers per industrial hectare. They aimed to cover only about one-third of each individual plot with the industrial building so that there was space to double the size of the covered area on the same plot without undue overcrowding.

The amount of covered space per worker in the first batch of new towns was analysed in the early 1960s and this analysis produced an average figure of 18 m². Welwyn was the exception at 27 m² and a survey conducted in Ipswich in 1965 produced a similar figure.

TYPES OF INDUSTRIAL ESTATE

The private sector and the public sector have much in common when an industrial estate is being planned but there are sufficient differences to justify separate study. The public sector should again be broken down as follows.

Large Industrial Estates

These are estates built in areas of high unemployment by a corporation set up by the Government for this purpose. Their sole purpose is to attract additional industry to the area. A scheme for such an estate on Teesside was prepared by Lord Holford. It has not been a success and most of the estate is still unoccupied. The reasons for its lack of success are probably economic and not due to faults in the planning. It is an interesting scheme and much can be learned from it. It is an attempt to introduce the Radburn concept, normally associated with residential development, into the layout of groups of factories. Teesside is examined in some detail at the end of this chapter.

New and Expanding Towns

Many of the considerations which have to be taken into account when an industrial estate is being planned in a new town also apply to an expanding town. New and expanding towns can therefore be considered together although the financial implications are different. They provide alternative means for the relocation of overspill population and industry arising out of the redevelopment of large urban areas after the Second World War. It was anticipated that substantial numbers of the original population of cities could not be accommodated in the redevelopment schemes envisaged and that they would need homes and employment in new localities.

The Government has no statutory power to compel industries to move to new or expanding towns although the granting or the withholding of Industrial Development Certificates may be used as pressure. Both new and expanding towns must attract industry by whatever means they have at their disposal. Powers of compulsory purchase enable them to acquire land at prices considerably lower than those for industrial land in established areas, resulting in lower rents. They are also able to offer new subsidised housing to key workers of industries moving in and there may be tax advantages and other financial assistance to attract industry. Expanding towns have a slight advantage over new towns since they can usually expect help from the exporting authority that has joined forces with them to promote the expansion scheme. There will be nonconforming industries which the exporting authority are anxious to move out of their present premises and they will often be prepared to offer inducements to encourage these industries to move.

It should be noted in this connection that there are indications of a change in attitude by some exporting authorities, who are now considering retaining industries in central areas to provide employment—there is feeling among many planners that large-scale relocation may have been overdone.

TYPES OF FACTORY

The types of factory likely to be in demand on an industrial estate in a new or expanding town can be classified into three main groups.

Purpose-built Factories

There will be a limited demand for industrial land for building a factory to meet the special requirements of an industrialist. It occurs when an old-established factory, usually in the exporting authority's area, has been persuaded to move. This situation should be met with a flexible attitude and either a factory should be built to the industrialist's specifications, or a suitable plot should be made available on long lease for the industrialist to build to meet his own requirements. This type of occupier could be of considerable value to the town and should not be discouraged, but should not be allowed to dominate the estate. It may sometimes be better to offer exclusive occupation of a plot of land of suitable size adjacent to but not part of the estate. Figure 12.1 shows how a small group of factories or warehouses can be good neighbours with a large plant.

Standard Type of Factories

These are factory units of between 460 and 920 m^2 which can be adapted for use by many industries. They should be so sited that they can be occupied as single units or in pairs. This point is well illustrated in figure 12.2.

Small Nursery Factories or Workshops

These can be as small as 180 m^2 and they are suitable for many small businesses. Since Industrial Development Certificates are not normally required for this size of premises, they are easier to let but they do not often contribute much to the employment situation. They should be sited in groups and at least some should adjoin each other. One section of the estate should be allocated to this type of development and it should be close to the main entrance to the estate. In this way it may be possible to arrange for centralised canteen and other services, including heating.

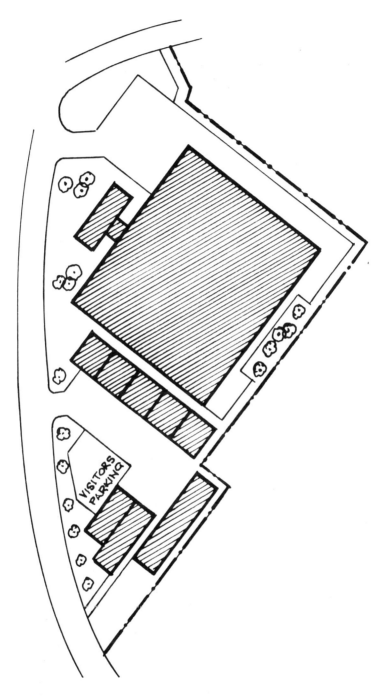

Figure 12.1 Small estate of light industry and warehousing

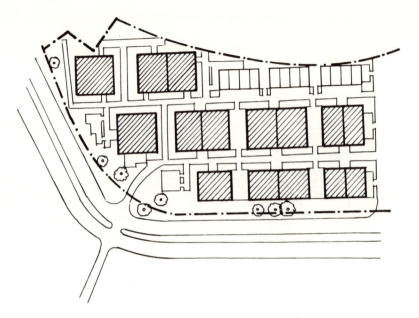

Figure 12.2   Compact industrial estate

OTHER LOCAL AUTHORITY DEVELOPMENT

Local authorities not concerned with an expansion scheme may still resolve to develop an industrial estate for a number of reasons. The first and most likely reason is that there are a number of factories located in a mainly residential area and the council has plans to redevelop the area and at the same time take steps to provide alternative premises for the industries they displace. This enables the local authority to retain the industry in the area and it may mean a reduced liability in respect of compensation for disturbance. The second likely reason is where the local authority has adopted a policy aimed at attracting more industry to the area. In these circumstances the local authority has to take its chance with other developers, such as new and expanding towns, which are all trying to attract industry.

PRIVATE DEVELOPMENT

A number of nationally known firms specialise in industrial development and some local developers will also seize the opportunity if there is an apparent demand in their locality and planning consent on a suitable site can be obtained. Since this is a normal business venture in a market with competitors who are often subsidised, it is essential to keep down costs by avoiding any wasteful development and also by maintaining the maximum flexibility so that no interested parties have to be turned away.

124

BASIC REQUIREMENTS OF AN INDUSTRIAL ESTATE

A number of basic requirements are common to most forms of industrial development.

(1) The site should be well drained and sufficiently level to allow for large areas of flat surface under a continuous roof. Sloping ground can be terraced but this takes away some of the flexibility in the size of units which can be offered.

(2) The chemical make-up of the soil should be acceptable and the loadbearing capacity adequate for whatever is likely to be built on it.

(3) The site should be of a size and shape capable of being divided into plots to give the maximum freedom to meet demands for any size of unit.

(4) It should if possible be on the leeward side of any residential development.

(5) There should be an adequate supply of water both for human consumption and for use in industrial processes.

(6) There should be sufficient electrical power available and also gas if it can be installed and supplied at reasonable cost.

(7) The GPO will be called on to meet a substantial demand for telephones and other communications services.

(8) Main drainage should be available including facilities for the disposal of trade effluent.

(9) The site should be served by an adequate road system capable of taking increased industrial traffic without undue danger. This road system should give direct access to the national trunk road system without having to pass through residential areas.

(10) The site should be served or be capable of being served by a public transport system.

(11) Although not considered an essential requirement, a railway siding is sometimes a useful asset.

(12) The site should be detached from but situated conveniently near residential areas so that labour can be drawn and there should be a sufficiently large pool of labour within reasonable travelling distance.

(13) Police and firefighting services should be within reasonable distance. (On a large estate it might be desirable to provide land for a substation for the fire service.)

(14) There should be shops and other facilities to meet the needs of the workforce during breaks. If these facilities are not available locally they may form part of the new development.

Space Requirements

In addition to building plots for individual factories the following space requirements must also be borne in mind.

Vehicular Circulation Sufficient land should be allocated to provide for roads of suitable width for the amount of traffic anticipated. They should give direct access to all the industrial plots on the estate and be linked with the local system of roads and streets. The road pattern should be designed to give the maximum flexibility in both the size and shape of individual plots.

Pedestrian Circulation   A system of footpaths should be provided alongside the vehicular routes. Paths should also be designed to provide short cuts to bus stops and car parks.

Parking   In addition to the parking of trade vehicles and executive cars, which will normally be within the curtilage of the factory premises, there will be a parking requirement for visitors. This should be conveniently situated and have adequate directional signs. There should also be parking facilities for workers' cars which must be strictly allocated. The amount to be provided and the ratio of spaces to workers can only be decided after local research into car ownership.

Landscaping and Planting   An ambitious landscaping and planting scheme programme may not be possible, but on the smaller estate there should be opportunities to plant trees to screen off unsightly parts of the estate, to provide small open spaces for people to sit during the lunch break. It may also be possible to plant trees along the grass verges. In cases where the site has to be terraced to provide level areas it is often possible to landscape and plant the slopes between the terraces. It may also be worth while providing a space for ballgames to be played during the lunch break. This area should be flat, with a tall fence, and be turfed or surfaced with a nonabrasive material.

Ancillary Buildings   These will depend entirely on what facilities are already available in the area and also on the size of the estate. Those planned for the Teesside estate examined at the end of this chapter show the scope for this aspect of a development.

PLANNING AN ESTATE

In planning the layout of the estate the space requirements given earlier must all be taken into account, but the two most important factors to be kept constantly in mind are the lack of uniformity in the size of units required and the wide diversity in the industrial processes carried out. This means that a developer should keep all his options open for as long as possible so that he can meet any reasonable demand. This essential flexibility can be created in the first place and be retained while the estate is taking shape only if the roads are in the right place to provide access to all the plots envisaged and at the same time permitting plot sizes to be made larger or smaller as demand reveals itself. This principle is illustrated in figure 12.3, which is a hypothetical example, and in figure 12.4, which is an existing industrial estate. While the long-sighted policy of new towns in leaving sufficient space on individual plots to allow for 100 per cent increase in floor space is accepted, the high level of values and the limitation on space oblige developers to design for higher site

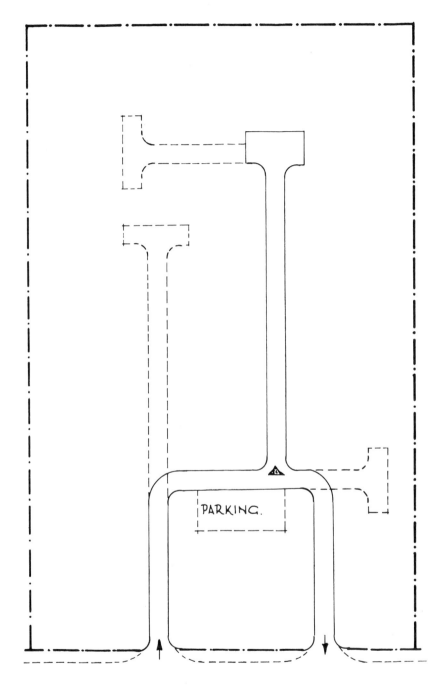

Figure 12.3 Road pattern on a small industrial estate designed to give maximum flexibility

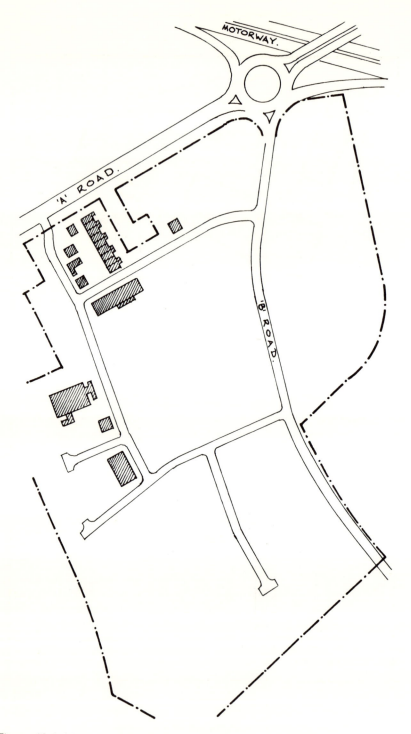

Figure 12.4   Large industrial estate immediately off a motorway

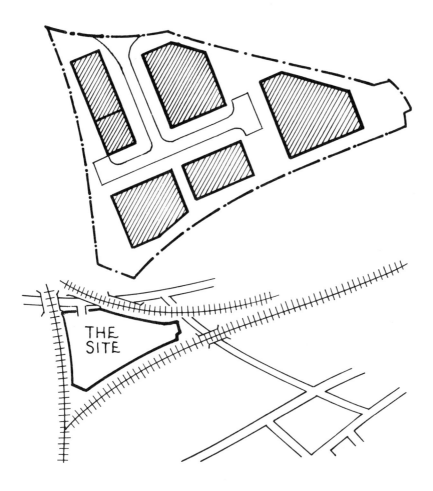

Figure 12.5   Industrial development on a restricted site

coverage; a good example of this is given in figure 12.5, which shows how six units can be placed on a restricted site. Figure 12.6 shows a dense but well-planned and orderly development with a wide range of sizes arranged so that two or more could be jointly occupied.

Siting the Buildings

Modern industrial estates will have a building line if the factories are to face directly on to a road. In any case the buildings should be set back to provide sufficient space for vehicles to drive in and manoeuvre without hindrance. There should also be provision for short-stay parking.

Where there is a generous allowance of site area for buildings to be extended

129

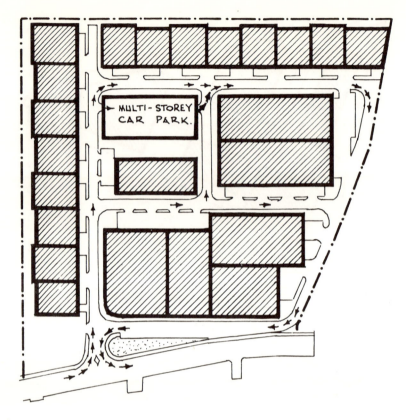

Figure 12.6   Industrial estate providing maximum option in sizes

at a later date, it is important to site the initial building with future extensions in mind. Figure 12.7 shows how a building, if originally constructed to one side of the plot, will allow for a second building of the same dimensions to be added later. If the first building is sited in the middle of the plot it may only be possible to build lean-to extensions on either side of the existing building, with a consequent reduction in efficiency of space usage.

INDUSTRIAL BUILDINGS

The detailed design of industrial buildings is a study in its own right, but in considering the site layout of an industrial estate an appreciation of the following basic components of an industrial building is essential.

Offices

If these are just for management use and administration of the business

130

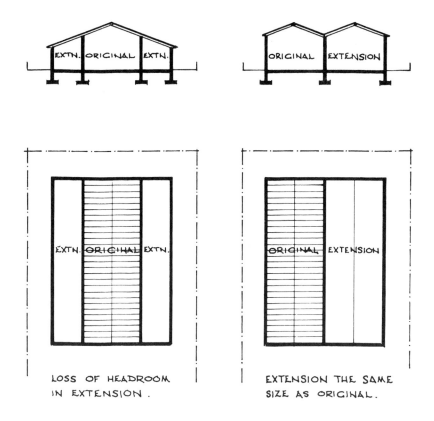

Figure 12.7   Alternative methods of extending a factory

conducted in the factory, they are not likely to be very large and can usually be accommodated on the upper floor of a two-storey office block at the front of the building. This then leaves the ground floor free for other purposes such as the provision of a canteen and rest room.

Production Area

This is the part of the building where the industrial process takes place. The amount of floor space allocated for this purpose will depend on the type of industry—it must have a direct relationship to the capacity to store both the raw materials and the finished product.

Storage of Raw Materials

Space for this purpose is usually allocated in the main building as near to the production area as possible. Materials not affected by weather or not likely to be stolen can sometimes be stored outside.

Stores for the Finished Product

Space allocated for this purpose should be as near to the point of despatch as possible to save double handling. In the manufacture of bulky or heavy goods the stores should be on the same floor level as the production area, with no obstacles between.

Receipt and Despatch of Goods

In a small factory it may be possible to provide one loading bay for both the receipt of raw materials and the dispatch of the finished product. The bay should be part of the main building so that it is protected from the weather and secure against pilfering. It should be near the main entrance to facilitate free movement to and from the premises.

Canteen, Rest Room and Toilet Facilities

These important parts of a factory should be sited away from noise and the canteen should have a pleasant outlook if possible. As suggested earlier the ground floor of the office block is suitable if it is available. The size of the canteen and the number of toilets and other facilities provided will depend on the number and the sex of the employees.

Ancillary Buildings

Certain materials used in industry must not be stored in the building that contains the production area. Additional buildings should be so sited that they do not take up too much open storage space.

INDUSTRIAL DEVELOPMENT IN INNER CITY AREAS

The fact that there has been much industrial movement out of inner city areas in the past 25 years can be attributed to one or all of the following reasons.

(1) Lack of freedom to use the industrial premises to the full because of restrictions imposed by planning control where the occupation is nonconforming.
(2) The use by the Government of Industrial Development Certificate procedure to deny a move to one locality but encourage a move to another.
(3) Grants and other financial inducements that are made available to industrialists prepared to move to selected areas.

A substantial change of emphasis is now taking place in an effort to close the void which has been caused by factories closing and industry and their workers moving out with nothing economically viable taking their place. A Government policy statement made in mid-1977 gave considerable weight to this change in emphasis. The statement outlined a number of major steps to be taken which can be summarised as follows.

(1) A new priority to be given in the main Government policies and programmes so as to provide a better life for people in these inner city areas.
(2) A more unified approach to urban problems (DOE to take over responsibility for the urban programme).
(3) Immediate priority to be given to strengthen the economies in these areas, for example, the needs of industry, small firms, etc.
(4) Policies on population movement to be reviewed and changed.
(5) The urban programme to be recast to cover economic and environmental as well as social projects.

The remainder of this chapter is devoted to a review of a plan for a large industrial estate in north-east England.

TEESSIDE INDUSTRIAL ESTATE

The following is a digest and review of the plan rather than a report on the development.

The Teesside plan is a very ambitious scheme for the development of a large industrial estate in the north-east of England. Unfortunately, the scheme has not grown as rapidly as was hoped, but this does not make it a bad scheme. Progress has been slow and, quite probably, disappointing to the Government, the local authorities, the planning team and not least to the people living in the locality who had good reason to hope for a substantial increase in the number and variety of jobs available.

The report and development plan,[4] published in 1964, were prepared by a project group under the chairmanship of Professor J. H. Napper. Lord Holford, who gained considerable experience in the planning of the Team Valley industrial estate in the 1930s, was appointed adviser. Some of the forecasting turned out to be wrong, but this can happen in any attempt at forward planning, particularly when it concerns human behaviour. Some of the views expressed may now be considered to be a little out-dated but the report contains much useful material for study in this specialist field of planning, as considered under the following headings.

Regional Situation

The location of the site in relation to adjacent towns and the population of those towns are important factors to be taken into account when planning an industrial estate of the dimensions proposed. The way in which the site fits into the national road and rail systems is considered, along with the facilities offered for sea and air travel.

The Site

The site was to have a final employment capacity of 10 000 jobs. It totals about 136 hectares and is rural in character, surrounded by good countryside. The site is undulating, with falls of 1 in 25 in some parts and almost level in others. The planting on the actual site is restricted to hedgerows and rather poor quality scattered trees. Climatically the site is exposed and open to winds from all directions. It is in an area which enjoys drier and sunnier weather than much of the north of England. The prevailing winds are westerly, although the most severe weather is usually associated with north to east winds. There are no mining operations in the area, nor are there any other apparent geological complications. The topsoil is unusually deep in some places. In this section of the report details of existing services are also given.

Factors for the Future

This section deals with the difficult problem of estimating how much expansion space to allow for each factory. Evidence of growth available to the group varied from industry to industry, and in some cases it also varied within an industry, so that no common factor could be found on which to make an assessment. The problem was made more difficult by the group's inability to predict the types of industry that would occupy factories they were planning. To get round this problem, a general policy formulated to plan sites for factories with only 100 per cent expansion space immediately adjacent. In some special cases this could be increased by spreading over into land originally allocated to an adjoining factory if that neighbour could see no future need for it. Outside the general policy, consideration would be given to special cases where there was evidence from the start that expansion was likely to exceed 100 per cent in a very short time.

It was thought likely that some types of industry would be more likely to expand than others, and so the industries likely to be attracted to a new industrial estate were broken down into two groups: *Group A*, industries increasingly affected by mechanisation and therefore unlikely to produce high employment figures per covered acre; *Group B*, those expected to require additional space as productivity increases. Group A industries are print, paper, publishing, timber, furniture, engineering, electrical goods, building materials, drink, tobacco, pottery and glass. Group B industries are services, other manufacturing, metal goods, leather, clothing, nonferrous metals, construction, distributive trades, textiles and food processing.

The possibility of siting a number of factories for similar industries or processes together was considered, but was abandoned after discussions with employers, who thought that this might accelerate the turnover of workers in similar trades. It was also felt that it would have a generally unsettling effect on attitudes to wage levels and working conditions.

In considering the amount of flexibility to allow for in the size of production areas, the group saw no reason for changing the principles adopted on other estates in the north-east—to encourage light and medium industries that do not initially require large production areas. Teesside already has its large-scale industry and there is therefore a need to diversify. This also meant that some factory units could be built in anticipation of demand.

With all the above facts in mind sites were planned with a depth of either 90 m or 120 m.

The report also deals with possible changes in the demand for manpower arising out of increased automation and with the special public transport needs of shift-workers. In considering the provision of public transport and parking facilities on the site the group referred to the Team Valley estate. It was estimated that the 15 000 employed on the estate came to work as follows

Bus during normal peak hours	6600
Bus outside normal peak hours	1000
Private car or motor cycle	3400
Bicycle or walking	4000

On the basis of these figures, modified to meet known trends, it was estimated that there would ultimately be a need to provide for 4000 private cars on the Teesside estate.

The possibility of providing a separated system of cycle track was considered but it was decided that this would not be justified owing to current evidence of a substantial reduction in this form of transport.

In considering the question of the separation of pedestrians from vehicles in the light of the expected dramatic increase in the latter, the group fully subscribed to the opinion that the motorcar must be controlled and an environment sympathetic to simple human needs and demands must be created. The following quotation from the report adequately expresses their view

'Trading Estates are not merely areas for the production of finished goods but are also environments where people spend a large proportion of their working life. As such we consider that the Estate should reflect the increased standards of environment demanded by a civilised society and as part of the upgrading of the industrial north and in particular the North East, we recommend that only the highest standards of design and planning should be accepted at every stage of the development.'

The group adopted a limited Radburn type of system, where vehicles are separated at the same level. They found no evidence to justify a completely separate system and therefore decided that the pavements should be alongside the roads. In the ordinary way there is little pedestrian traffic from factory to factory or even from block to block. The plan therefore concentrates on pedestrian routes from car parks and bus stops to factory groups independent of the main service roads. There are also car-free areas in the factory blocks. These can be seen in the plan of a superblock in figure 12.8.

The adoption of the Radburn concept in planning an industrial estate means that the approach by road is to that part of the factory which in conventional development would be at the back and is usually the most untidy. Radburn therefore demands a higher standard of overall design.

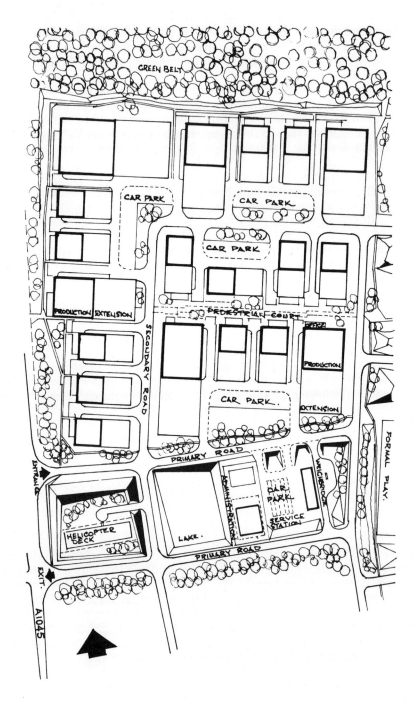

Figure 12.8   Teesside industrial estate: industrial superblock

There are other signs of a move away from the old concept of the prestige office frontage with a light steel-frame structure behind in favour of one coordinated building housing all the work activities.

Consideration was given to leisure activities. It is possible that the decrease in working hours will result in an increase in the demand for organised games on the estate and space has been allocated for this purpose. There are also areas for a kick-around at lunch time, and other areas just for walking or sitting in the sun in good weather.

The Development Plan

This section again emphasises the importance of maximum flexibility, as the following quotation shows

> 'The development plan must not only be able to accept and exploit the enormous differences from factory to factory but must also be capable of meeting the changing demands of industry in general as the estate grows and develops.'

To assist in planning individual sites, the group looked at the elements which normally go to make up a factory. These were office accommodation, canteen and social room, covered production area, covered storage area, open storage area, ancillary buildings, expansion space, vehicular access, car parking and pedestrian access. They emphasised that the proportion and order of priority vary from factory to factory and they dealt with each in turn in some detail.

Office Blocks   The siting of the office blocks away from the estate road was the first move in the acceptance of the Radburn concept. They are located in a central mall and form vehicle-free office courts, as shown in figure 12.9. The mall is a 30 m wide strip between the actual factory buildings. The range and sizes required by individual factories should provide contrast and variety in the office courts. The office courts were to be simply landscaped with paving grass and trees.

Factory Canteen and Social Rooms   The possibility of centralised catering was considered but it was found to be unpopular with workers. Each factory should have its own canteen and social rooms, to form part of the main factory structure, and if possible with windows looking out on to the office courts.

Covered Production Areas   Maximum flexibility is essential in these areas. Each production process would impose its own demands and limitations but the constant requirements of vehicular access, pedestrian routes and service facilities apply some discipline to the design.

Covered Storage Areas   Requirements vary greatly between one industry and another and are generally closely related to the size of the production area.

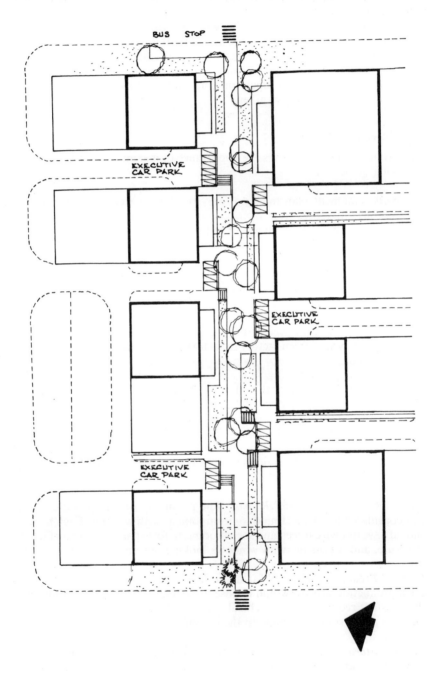

Figure 12.9   Teesside industrial estate: pedestrian office court

138

Screened Yard for External Storage The question of open storage was considered by the group, knowing that this is invariably the most untidy part of most industrial premises. They did not consider it practicable to expect the occupier to keep all the plant and machinery associated with production within the actual factory building, but recommended that when an open storage area was provided adjacent to the factory it should be screened from general view by some form of palisade.

Ancillary Buildings These should whenever possible be sited as an integral part of the main structure.

Expansion Space This was considered earlier in the report and was the subject of a policy decision limiting the increase in size to 100 per cent. The group anticipated that the initial development would consist of the office block and the factory block. Since this would occupy the full width of the plot, the expansion space could only be close to and visible from the estate roads. It would therefore require screening in the same way as the open storage areas. (See 'Landscaping', p. 141.)

Vehicular Access and Parking Individual factories require three categories of vehicular access off the secondary roads.

(1) Industrial traffic consisting of lorries and vans needs direct access to the factory. This is via a factory service road 6.6 m wide situated within a strip of land 15 m wide between each factory and the next.
(2) Executive cars related to office blocks also need to use the factory service road, with parking facilities at the head close to the office block and screened from the office courts.
(3) General parking areas are off the secondary roads and allocated to a group of factories. These have been left as flexible as possible, with parks at ground level only in the first instance but provision has been made for a second-tier deck over the original site if demand justifies it.

Pedestrian Access Special arrangements have been made to reserve a pedestrian route from the site of a large proposed residential development to the estate since it is anticipated that many of the new residents will work on the estate.

The Road System

Access to the estate is off an A category dual-carriageway road (A1045). The internal road system, shown in figure 12.10, is as follows.

(1) A primary network consisting of two carriageways, each 7.2 m wide, with exceptionally wide central reservations to accommodate playing fields and other recreational areas. The roads will also have laybys for stationary vehicles.
(2) A secondary system made up of roads each 9 m wide carrying traffic in both directions to serve the needs of the factory superblocks.
(3) Factory service roads within the superblocks (see p. 136).

139

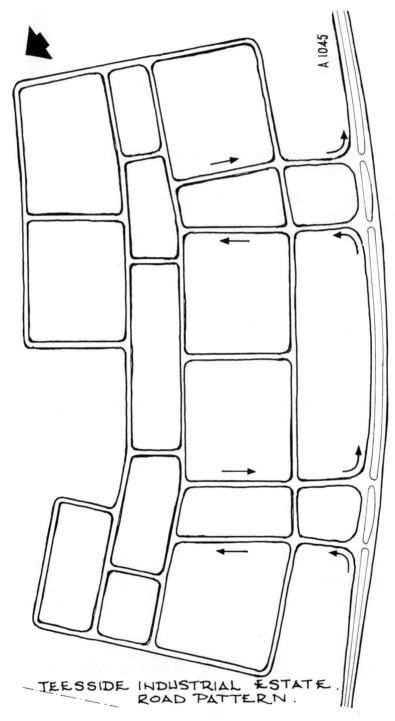

TEESSIDE INDUSTRIAL ESTATE.
ROAD PATTERN.

A 1045

Figure 12.10   Teesside industrial estate: road pattern

Superblocks

This word is used in the same way as in a Radburn residential layout to describe an area in a development formed and enclosed by primary roads. The report contains the detailed layout of one superblock, reproduced in figure 12.8. Eight factories are grouped around the pedestrian mall. Because the superblock occupies a corner site, factories are on the perimeter, with four to the north facing the green belt and seven to the west facing the planted shelter belt of trees between the development and the main road that runs alongside the site. There are car parks and some of the estate buildings are located in the block.

Estate Buildings

In addition to the factories and their associated offices the scheme includes the following.

(1) A central administrative building to be sited conveniently near the main entrance. This is primarily to accommodate those concerned with the management of the estate but the group suggested that this block could ultimately include
   (a) executive suites and board rooms to be hired by tenants for specific occasions
   (b) executive dining rooms and club rooms
   (c) residential penthouses for visiting executives
   (d) personnel and welfare services, clinic, nursery, etc.
   (e) central computer service
   (f) office services, typing, duplicating, printing, etc.
   The group considered that only the highest standards of design and construction should be accepted for this key building and its landscaping. The car parking associated with the centre is to be sunk below ground level and further screened with earthworks.
(2) The estate petrol and car service station is to be sited on the primary circuit road.
(3) Changing accommodation with additional car parking is to be sited in the recreational spine formed in the wide reservation between the two carriageways of the primary road system.
(4) The provision of a fire station was considered but the local chief fire officer advised that existing facilities were adequate.
(5) An information office is to be sited on the 'in' road of the primary circuit.
(6) A weighbridge is to be located near the service station and made accessible to both incoming and outgoing traffic.

Landscaping

The proposals for landscaping are as follows.

(1) A permanent shelter belt of trees is to be established and maintained as commercial forestry with the prime purposes of providing

141

(a) a wind break

(b) a background of wooded landscape for the building development

(c) a financial asset.

(2) Existing woodland is to be treated by thinning out and transplanting hardwood.

(3) Playing fields are to be laid out in the central reservation.

(4) There is to be strong earth modelling by the use of the spoil from the earlier levelling operations.

(5) Planting is to be carried out on the periphery of superblocks and along secondary roads to provide screening to factories and car parks.

(6) Amenity planting of trees, shrubs and grass is to be carried out in the pedestrian courts. (A special note is made with regard to the responsibility for ongoing care and maintenance after the initial planting.)

(7) There is provision for rough grassing down of areas scheduled for the later stages of the development.

(8) A tree nursery is to be established, together with accommodation for maintenance staff.

The final section of the report deals with the services and includes the source and availability of the usual services. District heating is considered as a possibility and discussions took place with the representatives of both the gas board and the National Coal Board. Gas was ruled out but the National Coal Board undertook to conduct a feasibility study. Oil companies were also contacted.

The report contains much valuable information for a student of planning and development. Some people may quarrel with the way in which some of the factual information has been interpreted and with some of the conclusions drawn but this does not detract from the value of the report as a well-researched, well-thought out and well-presented document. That is where its real value lies and that is the reason for its inclusion in this book.

REFERENCES

1. *Report of the Royal Commission on the Distribution of the Industrial Population*, Chairman Sir Anderson Montague-Barlow (HMSO, 1940)
2. *Report of the Committee on Land Utilisation in Rural Areas*, Chairman Sir Leslie Scott (HMSO, 1942)
3. Patrick Abercrombie, *Planning and Reconstruction* (Todd, London, 1944)
4. J. H. Napper, *Report and Development Plan for Teesside Industrial Estate* (Board of Trade and Industrial Estates Management Corporation, London, 1964)

13 SERVICES

The many services a developer will be concerned with can be subdivided into the following main groups: incoming services, drainage and waste disposal, public transport, other general services and specialist services. In this chapter each of these groups will be considered in turn.

INCOMING SERVICES

Water

The provision of water throughout the United Kingdom is now the responsibility of 10 regional water authorities, set up under the Water Act, 1973. Prior to that date water was supplied by local authorities and in some cases by water companies. Some of these water companies remain but they are likely to be absorbed by the new regional water authorities quite soon. A system of direct billing is now being introduced. Until such time as this is completed water authorities have the power to make a precept on local authorities, requiring them to collect water rates on their behalf. This is based on the rate poundage and is levied at the same time as the general rate.

The supply authority is normally responsible for the main service pipes. These are usually laid under the grass verge or, in the absence of grass verges, under the pavement, and are often on both sides of the road. They are laid deeply enough to avoid frost (see figure 13.1). The responsibility of individual occupiers begins at a stopcock, usually located just inside the frontage boundary.

As was mentioned in chapter 3, preliminary investigations should be made to establish that water will be available in sufficient quantities. This is particularly important in the case of industrial development when large quantities of additional water are required for industrial processes. The water may be potable (drinking water) or it may be drawn from a local river or stream, but in all cases the supply will be controlled by the water authority.

The demand for a supply of water can be calculated on the basis that the daily supply for a residential town is likely to average 135 litres per head of population and in an industrial town up to 230 litres.

Electricity

Electricity is produced by generating boards and fed into a national grid. Regional electricity boards are supply authorities and they are responsible for the provision of electricity to domestic and other consumers; most of the power supplied is at 230 to 250 V a.c. In urban areas the supply is carried

NARROW FOOTWAY.

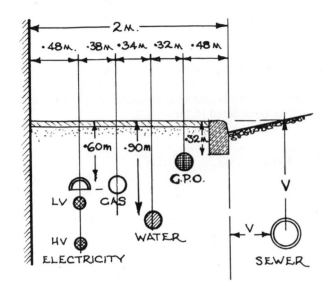

WIDE FOOTWAY.

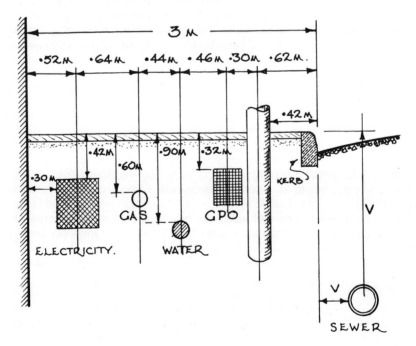

Figure 13.1   Services below ground

underground (see figure 13.1), but in rural areas it may still be carried by untidy-looking overhead cables. Financial aid from central government or local funds may be available for moving the electricity cables underground in particularly attractive or designated conservation areas.

The supply authority usually installs a service panel inside the building of the property to be supplied and the responsibility for individual wiring is with the contractor. However, there is a very rigid safety code to be complied with before the supply is connected and the authority has wide powers of inspection. When the service is underground the cable is laid by the service engineer employed by the supply authority. There must therefore be constant liaison between the main contractor and the authority so that all site works can be combined and carried out in the right sequence. The supply authority will usually approach the developer for a contribution towards the cost of the work entailed and in certain circumstances may demand the full cost. This will normally be a matter for negotiation and will often depend on the anticipated consumption. There is invariably a charge for connection. The supply of electricity to an industrial estate will depend largely on the particular industries and their anticipated consumption. Voltage requirements may differ and the developer should be ready to allocate land for a transformer station if asked to do so.

Gas

Gas is supplied by regional gas boards, with the National Gas Council coordinating supplies and giving guidance. Widespread changes have taken place in the gas industry in recent years following the discovery of natural gas in the North Sea, and it is probable that all gas supplies will come from that source until it is exhausted. Before the discovery of natural gas, coal was traditionally the raw material for town gas production, but it was being largely replaced by volatile hydrocarbons from oil and imported natural gas. Until that time the appliances originally designed for coal gas could still be used but the chemical content of natural gas is so different that appliances have to be either specially made or adapted.

Gas is normally available in the old-established urban areas, but not in rural areas. The initial cost of installing a gas supply in new areas is high, and may not therefore be financially justified. However, the availability of gas to an industrial estate will widen the field of possible occupants. A number of industries consume large amounts of this important source of industrial energy.

In the case of a residential development, if gas is available in the locality, there is likely to be a demand for it. Many people prefer to cook by gas and gas-fired central heating if often preferred to solid fuel or oil. There is also the additional advantage that no storage tanks are required and this is particularly important in high density areas. Supply authorities are responsible for the underground pipes up to and including a meter in the buildings to be supplied (see figure 13.1). From the meter onwards the responsibility is the consumer's but the authority has the duty to inspect new work and to carry out tests when making the connection; a charge is made for this service. It is most important that any pipelaying should be carried out as part of a detailed programme of site work. The authority has the power to make a charge for this work and the actual amount is a matter for negotiation and will, as with electricity, usually be related to the amount of gas that is likely to be consumed.

GPO Telephones

The telephone has become part of the modern way of life and the telephone service is considered essential to all forms of development. Early contact with the area telephone manager should ensure adequate provision, particularly if an estimate can be prepared showing the likely number of users. The GPO still has a monopoly on almost all forms of telephone communications and is fully equipped to provide a service, which can include line and radio communications, links required by firms using centralised computers, teleprinters, close-circuit television and other sophisticated forms of electronic equipment.

A developer will not generally be involved in the detailed arrangements between the GPO and the user, but he should make certain that the service is available and that any cablelaying in addition to wiring is carried out as part of the co-ordinated programme of site works. Co-operation with the GPO in providing a site for a telephone call box will enable it to be properly sited in the first instance, rather than added as an afterthought when public demand arises.

Street names and arranging house numbers are matters for the local authority, but any way in which a developer can assist in the easy identification of occupiers will help the postal services of the GPO.

DRAINAGE AND WASTE DISPOSAL

The regional water authority has overall responsibility for most aspects of drainage but a developer will normally make his contacts through the local authority because it has agency arrangements to deal with everyday matters in its exercise of building control and the Building Regulations. A natural system of drainage operates the whole time and, although this is quite unable to meet the drainage needs of modern society, it should be kept in working order.

Land Drainage

It is sometimes necessary to increase the efficiency of the natural land drainage system in low-lying areas where the level of saturation (water table) is high. This can be caused by a fault in the stratified formation of the subsoil or by something as simple as a disturbance in the subsoil caused by the backfilling of a deep trench or even the foundations of a new building. This situation can often be corrected by putting in land drains to permit the underground water to follow a natural course. In rural areas land drainage is important to agriculture and therefore new developments should not interfere with the drains formed by ditches and streams passing through the site. Streams should be culverted or diverted and ditches should not be filled without first laying down land drains of adequate size.

Sewage Disposal

The local authority has a duty under public health legislation to provide such public sewers as will be necessary for the effectual drainage of their area. It also has the power to require the connection of existing premises to the sewer and to lay down standards in accordance with the Building Regulations. The mainly liquid refuse which goes into the sewer comes from three main sources

Domestic sewage
Surface water
Trade effluent.

The volume to be dealt with varies at different times of the day. The two peak periods for domestic sewage are usually early in the morning and late in the afternoon. Surface rainwater will depend on the rainfall, and trade effluent can be a regular flow or it can be in large quantities at regular or irregular intervals. Domestic sewage can be calculated on the simple basis that the quantity of liquid that goes into a building is likely to come out later through the drains into the sewer. The amount of rainwater entering the sewer can be controlled to some extent by keeping the rainfall that runs off the roofs of buildings separate from the rainfall that runs off paved areas, and by disposing of the two kinds of rainfall by two different methods. These methods are described in more detail later in this chapter. The local authority has special statutory powers to control trade effluent.

In a new development requiring the extension or alteration of an existing sewage system the normal practice is for the work to be carried out by the developer and that part of the work which falls within the definition of public sewer is taken over by the local authority, following a procedure similar to that for roads referred to in chapter 7.

If the local authority requires additional work in the way of larger pipes, different materials, depth, gradient or direction in order to include other developments in the scheme, they may specify these additional requirements and will be liable to meet any increase in costs arising out of the work. Differences may be settled by arbitration or in the magistrates' court. The structural details of the sewer will be subject to approval by the local authority as part of the Building Regulations submission and should conform to the particular requirements of the system operating in the area in respect of what goes into the system and what is disposed of by other means. The system most commonly used is that known as the 'partially separate'. This means that only the domestic waste from the kitchen, bathroom and W.C. and that part of the rainwater that runs off the roofs of buildings goes into the sewer, the rest of the rainwater being diverted into a different system, often known as a stormwater sewer, which can then flow into the nearest watercourse because it is unpolluted by human waste. A combined system with all the surface water running into the sewer will be found in some old-established urban areas. The separate system is that found in mainly rural areas, where sewage is raised by pump to a level from which it can flow by gravity to the next pumping station or to the disposal works. The capacity of the pump and the size of the pipes usually means that the load on the system has to be controlled. For this reason only the foul waste goes into the sewage system and rainwater is soaked away or diverted to the nearest watercourse.

There are a number of basic principles to be observed in the design of a foul drainage system.

(1) Pipes should be round and true to shape, hard and of adequate strength and with a smooth and impervious inner surface.
(2) They should be laid with a fall (gradient) to permit a flow sufficient to make the system self-cleansing.
(3) They should not be overlarge but should be adequate to take the peak load.
(4) They should be properly supported throughout their length.
(5) They should when possible be laid at a sufficient depth to prevent crushing; alternatively they should be set in concrete.
(6) They should be laid in straight lines and the fall should be constant between points of access.
(7) Joints between pipes should be flexible but watertight.
(8) Branch drains should be kept as short as possible.
(9) The system should be adequately ventilated and inlets fitted with traps where appropriate.
(10) Manholes for inspection and cleaning should be constructed to make the system accessible at the following points
  (a) the head of each branch of the system
  (b) bends and junctions
  (c) changes of gradient
  (d) intervals of not more than 75 m in straight runs.
(11) Only under exceptional circumstances should drains pass under buildings.

The sewers are laid either under the carriageway or under the grass verge or pavement alongside the carriageway. This is convenient if the system is a combined one because the water from the roadside gullies can run directly into the sewage system. It also means that the pipes are accessible if they are in need of repair, although of course it is appreciated that this may be at the cost of some inconvenience to the public. These factors should be taken into account when carrying out the site planning of an estate, but they should not be followed as unbreakable rules. There may be good reasons for divorcing the sewage system from the road pattern but in this case pipes should not be laid under buildings.

Trade Effluent

In planning the development of an industrial estate, the disposal of liquid waste which results from industrial processes must be given special consideration and discussion must be held with the local authority. The starting point in any discussion will be that it is generally illegal to discharge into the public sewer any substance which is likely to damage it.

The local authority may grant permission for the discharge of such trade effluent but it has the power to impose conditions in respect of temperature, volume and the acid or alkali content and to require the neutralisation or the elimination of any constituents that are likely to make treatment of the sewage more difficult. The authority can require meters to be fitted to monitor the input; it can take samples and demand payment for the service.

Refuse Disposal

Domestic refuse is collected by the local authority, usually on a weekly basis. A developer will be concerned with access for this purpose and, in the case of the design of individual houses, with providing a suitable place for the bin to stand. If the bins are mechanically handled they will need to be of standard design; these are usually provided by the local authority, which may make a charge for them although it has the power to provide them free. The storage of refuse in nonreturnable sacks is becoming more popular because it is more hygienic. In blocks of flats of four or more storeys a common solution is for individual bins to be emptied down a refuse chute into large containers. In designing the chutes such factors as noise, smell, fire risk, blockage and the overflow of containers must be taken into account. The container chamber should be kept free from vermin, flies, wasps and the like, and good ventilation at all access points in the system is essential. The vehicle collecting the containers should be able to back into the chamber to pick them up.

Commercial and trade refuse includes domestic refuse but also includes byproducts of business. Often included are large quantities of waste paper. Some businesses prefer to make their own arrangements, particularly if their waste paper contains matters of a confidential nature. However, the local authority must collect from hotels and restaurants but it has the power to make a charge. Refuse from industry must be considered independently. Owing to its special character some of it will be similar to domestic or commercial waste, but much of it will be waste from industrial processes and this could be dangerous or offensive. Some of it may have a salvage value. The best arrangement is for the local authority to collect the domestic type of refuse and for the industrialist to make separate arrangements with a contractor to collect the remainder. In all cases provision for proper storage before collection must be made and this must have adequate vehicular access.

PUBLIC TRANSPORT

Most residential estates and all industrial estates and town centres are likely to benefit by being served by a local bus service. The developer can often encourage the local operators to extend or change their route by collaborating with the local highway authority in providing facilities for the service. This may include widening estate roads to provide for bus stops and, if a residential estate is likely to be the furthest point of the service, by providing adequate turning space and possibly also a layby for a bus temporarily out of service. The developer may also like to make a gesture of goodwill by providing a covered bus shelter.

OTHER GENERAL SERVICES

These are services mainly concerned with safety and security.

Street Lighting

Street lighting is traditionally associated with security and is still one of the public's best forms of protection against thieves and vandals. It should also be provided for the convenience of road users, both vehicular and pedestrian, and for local residents.

The provision of street lighting in a new development is usually a matter of co-operation between the developer, the local authority and, usually, the local electricity board. Gas is still used for street lighting but in most new developments electricity is of course used. The installation, siting and height and types of lantern are a matter for lighting engineers, and are covered by British Standard Codes of Practice, but the developer will be concerned generally and should therefore have some knowledge of the subject.

Types of Street Lighting   The different types of street lighting are classified into groups. Group A and its subgroups are concerned with main traffic routes and C,D,E and F are special forms of lighting, such as that for bridges, tunnels and the like. Those with which a developer is most likely to be concerned are group B, which covers ordinary street lighting, and G, which deals with town and city centres. There is one further category, which can hardly be classified as a group. This is known as 'police and amenity lighting', and consists of isolated lamps placed at strategic points. Lighting systems in group B are used when it is considered by the local authority that an area requires adequate lighting and it normally applies to residential streets and footpaths and some of the roads and streets in minor centres and other areas.

Lighting should be adequate and kerbs defined; there should be no glare and shadows should be avoided; the standard of light should be satisfactory from the police point of view, particularly in forecourts and the lower parts of building frontages. The column to support the lantern will normally be erected up to within 0.6 m of the carriageway, either close to the edge of the footway or on the grass verge. It might sometimes be possible to provide small recesses in the frontage of the boundary fence. The siting of the standards and columns and the distance between them and also the level of output will depend on the road formation and the amount of traffic using the road. They will normally be staggered, but they may be on one side of the road only on outside bends or they may be concentrated to cast light on to an important feature on the opposite side of the road. In a very narrow road where there is an avenue of trees the height of the standards will normally be 5 m but this may be increased to 6 m for wide carriageways. The normal minimum spacing will be 60 m but

150

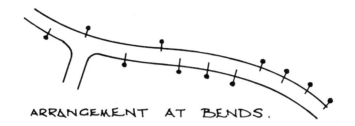

ARRANGEMENT AT BENDS.

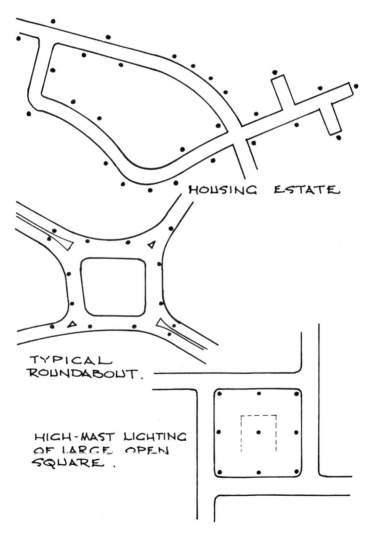

HOUSING ESTATE

TYPICAL ROUNDABOUT.

HIGH-MAST LIGHTING OF LARGE OPEN SQUARE.

Figure 13.2 Street lighting: the positioning of lamp standards

151

this can be extended on clear straight roads. Figure 13.2 shows examples of the positioning of lamp standards in different situations.

Fire Services

Most of the requirements for fire prevention and fire fighting are covered by the Building Regulations, but when dealing with the site planning aspects of a scheme the developer should appreciate the need to give access to fire-fighting appliances so that they can get close enough to buildings to do their job properly. The provision of an adequate water supply and the siting of hydrants may also concern the developer.

SPECIAL SERVICES

These are services which are usually not considered to be essential to a development but if they are included they may make a substantial contribution to the success of the scheme.

Community Heating

This service is now being provided in a number of new residential estates. There is a particularly large one at Runcorn, a new town in Cheshire, and also one in the expanding city of Peterborough, Cambridgeshire. The system consists of a centralised boiler providing hot water for domestic purposes and for central heating to other buildings through a network of heavily insulated underground pipes, thus making individual heating systems unnecessary. This work should only be carried out under the direction of specialists, and the developer's main concern should be to see that the pipes are laid at the same time as other services, to cause minimum disruption. Other services are referred to in chapter 8 on residential development and in chapter 12 on industrial development.

14 LANDSCAPING

The provision of open amenity space and its treatment as part of a modern development must be seen in the right perspective—we are never likely to return to the days of Repton and Capability Brown, who were able to plan and form flowing landscapes over many acres of privately owned parkland. Occasional opportunities arise in the planning of public open space in a new or expanding town but, generally speaking, a developer will be concerned with the buildings and the treatment of the land around the buildings. The economics associated with property development generally result in building at maximum densities, which does not leave much land for open space uses. At the same time people are becoming more conscious of their surroundings and therefore more demanding in respect of their comfort and wellbeing, both inside and outside their homes and places of work.

Any new development should present a challenge to all those concerned in its planning but the treatment of open spaces calls for the maximum ingenuity and can make the difference between an uninteresting group of houses and a pleasing and attractive residential estate. It can also soften the harsh lines and hide some of the ugliness of an industrial estate.

The aim should be to supplement nature and to try to encourage natural growth and create a suitable environment where trees can grow to maturity, shrubs can thrive and flowers bloom. Although grass is one of the most natural and pleasing treatments for large areas, it cannot be used everywhere—there will be paths to be surfaced and other hard paved areas essential to the development. A paved area need not be made of large, uninteresting concrete slabs—there is a wide choice of paving stones and textured material available.

A landscaping and planting scheme will depend on many factors, including climate, orientation, altitude and soil quality. It will also depend very much on what the site has been used for prior to the proposed development. A new town scheme on 'white land' which has been used for agriculture will have the advantage of natural conditions for trees and shrubs to grow in. The topsoil may have to be moved around or stored for a period but there is likely to be an adequate supply available on site. These favourable conditions do not usually occur in an urban renewal scheme, so they must be contrived. This often means earth-moulding and importing suitable topsoil from outside sources. It may even include grubbing up old foundations and realigning the system of land drainage.

THE APPROACH TO DESIGN

The approach to the design problem will very much depend on what already exists on the site in the way of trees and shrubs. The solutions available to the developer are not necessarily alternatives and there may be situations where a scheme will be a combination of them all. However, they do serve as a check list and may indicate where immediate steps can be taken to reduce the inevitable time lag associated with growing plants. The preliminary inspection of the site outlined in chapter 5 should provide sufficient information for a first draft plan to be formulated. It is important that all those concerned with different aspects of the plan should be in close liaison. This is particularly important in respect of the location of the buildings and the landscaping. Ideally the buildings should form part of the landscaping so that each building has an appropriate setting, but this ideal will seldom be attained without some drastic changes in the present approach to land use generally.

Courses open to the Developer

The treatment of trees or any other features already on the site should be influenced by the following considerations.

To Retain   It should be almost a golden rule to retain all sound mature trees already growing on the site and it may be worth locating buildings so that this is possible. The technique of plotting the exact position of all existing trees and other important features on to a transparent overlay is explained in chapter 5. Arrangements should also be made for the preservation of the subsoil by carting it away and stocking it in a convenient place.

To Protect   Having decided to retain trees and other features it is essential to issue firm instructions for their protection.

To Improve   Existing trees may be healthy but they may have been neglected and be in need of attention. If this remedial work is done early it will enable a tree to grow to correct its shape and to overcome the effects of the treat-ment.

To Add To   It is most unlikely that a site for development will provide a ready made and suitable landscape. Some tree-planting will be required and the scheme may justify planting a number of mature trees. These are available from several sources, including the National Coal Board and the Forestry Commission. The trees, which can be upwards of 6 m tall and with a girth of up to 0.5 m, are dug up and moved on a specially constructed digger–transporter.

To Contrive   There are occasions where a site for development is entirely flat and completely devoid of interesting features and where there are no trees or other planting. In a case such as this the developer has the advantage of a completely cleared site, which allows much greater freedom in planning the layout. Landscaping is more difficult, however, and calls for a high degree of skill. In order to get the maximum advantage out of such a situation, the landscaping scheme should be planned early and any major earth-moulding carried out if possible before building work begins. It may also be possible to carry out parts of the tree-planting programme in areas where it will not hamper building work.

To Maintain   This important aspect of a development project is likely to be overlooked until it becomes a problem. This is particularly true of residential development. (The importance of the private developer being left with no responsibilities after all the houses are sold as has already been stressed.

A landscape is a living thing and as such it will be constantly changing. Trees will get bigger and may need to be topped or their branches may need to be lopped, or the trees may even become dangerous. The shrubs will need to be clipped and the bushes cut back and where there is grass this will need to be mown. Local authorities and new town corporations are in the fortunate position of having labour resources for this purpose but the private developer should make certain that all land is taken over by someone so that the owner of the land automatically becomes responsible for whatever grows on it. Much of the land will be sold in the form of house plots to private purchasers, but early agreements should be made for the local authority to take over the grass verges and any open amenity land.

USE AND PLANTING OF TREES AND SHRUBS

Trees in the landscape have many functions and the following are typical examples of their use

(1) to form a wind break and to provide shelter from dust and, to a limited extent, noise
(2) to form a visual barrier to give privacy and to hide unsightly features
(3) to channel attention directly towards a special feature or away from an uninteresting view
(4) to demarcate boundaries and to create external spaces within a site
(5) to relate buildings to each other and to the site
(6) to direct pedestrian flow
(7) to form a backcloth to an important building or piece of sculpture and to provide contrast in the texture, colour and form of buildings.

Shrubs, although much smaller, can generally be used for the same purposes as trees.

Characteristics of Trees

The colour and texture of the foliage and the bark of the tree are important when selecting a suitable variety of tree for a particular purpose. The pattern of growth and the silhouette against the open sky is often a significant factor in choosing a tree; for this reason trees are classified according to their shape. There are six basic shapes: broad, round, square, tapering, conical and columnar. A further group of trees is classified as 'picturesque', based on the configuration of the branch—the 'weeping' varieties are the best known.

It is essential that trees and shrubs should be of a species suited to the soil

155

climate and the atmosphere in which they are planted. Evergreens do not flourish in a smoky atmosphere because the pores of the leaves become choked with soot. The London Plane is particularly good in a polluted atmosphere because it sheds both its leaves and its bark each year. Some of the more common varieties of tree suitable for different locations are given in appendix G.

Health of Trees and Shrubs

All plant life needs a wide variety of chemicals to sustain life. The raw materials from which the chemicals are produced are drawn in through the roots and leaves, and energy is derived from the Sun. Most of the chemicals required are found in normal fertile soil but the balance may need to be adjusted by the addition of such important chemicals as potash, phosphates and nitrogen. In addition to the chemicals which provide the food a tree also needs

(1) an adequate but not excessive water supply
(2) light and air
(3) moisture in the air
(4) the correct temperature range
(5) good anchorage for its roots.

Trees in Paved Areas

These demand special attention since they are expected to survive in an unnatural setting. Specimens must be carefully selected to have a well-developed root system and a strong stem. Drainage is important and the water table should be below the ball of the root. They are planted in a hole filled with loam to which fertiliser has been added. Newly planted trees need copious watering when they are first planted and the ground around the tree should be kept well dug and free from weeds. When the trees are established the trunk can be surrounded by an iron grid and the paving stones continued up to it, provided the joints are kept open to allow the rain to percolate through and air to get to the roots.

If impervious paving is required, surface drainage can be linked to a watering system for the trees by forming a french drain under the roots and diverting rainwater to it rather than into the normal stormwater sewer.

Bushes as a Deterrent

Bushes such as holly and other prickly species can be used to discourage people from leaving the footpaths. This is often done where children would be tempted to take a short cut across a garden. The bushes are sometimes planted under ground-floor windows of blocks of flats to stop children playing too close to the buildings.

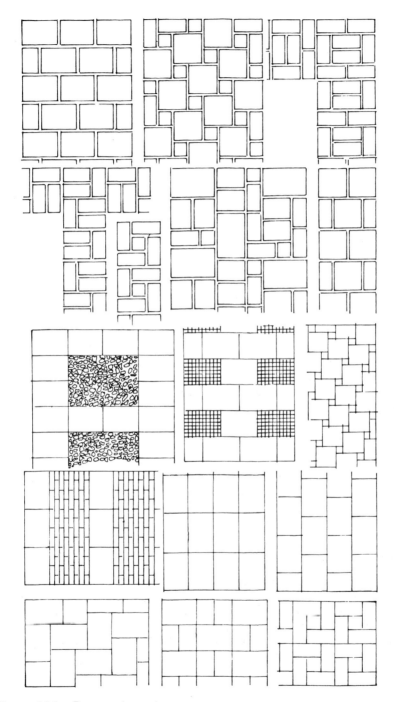

Figure 14.1   Patterns in paving stones

Grass

This is the most pleasant of all means of surfacing external surfaces in built-up areas. It provides for walking, sitting and children's play. When only a small grassed area can be provided it is important to raise it slightly above the paved areas. Grass needs light and air and will not generally grow under projecting balconies or under some types of tree.

Ground Creepers and Shrubs

The species of shrub known as prostrates or creepers, which cling close to the ground or other surfaces, can be used to advantage to cover unsightly spots and they will generally grow in shady areas and other places unsuitable for grass.

Hard Surfaces and Paved Areas

Hard surfaces are easy to drain and are therefore cleaner under foot in bad weather. Grass will not stand up to the hard wear caused by people constantly walking in one place. Paths, entrances and other places where people assemble must have a hard surface. Concrete is the most commonly used material for hard surfaces and it can be made interesting by colour and changes in texture.

Paving stones, if properly laid in the first instance, can provide a trouble-free paved surface which will last up to 30 years and will need little or no maintenance. Since they can be obtained in different sizes they can be laid to form a variety of interesting patterns, as illustrated in figure 14.1. They should be bedded on a compacted and stable base immune·from the effects of water and frost. The joints should be properly pointed. Concrete edgings should be provided to prevent water getting into the base. Granite setts provide a hardwearing surface but they are often a little uncomfortable to walk on. Cobbles can be used for parts of the paved surface to discourage people from walking on it. They can also be used to protect grass verges. Smooth nonabrasive asphalt has many advantages and can now be obtained in a variety of colours. It is used extensively for children's play areas. Rough, dark surfaces should be provided where cars are likely to stand, to break up oil stains.

Plant Containers

These can be used where it is difficult or impossible to plant trees and shrubs in the ground in the normal way, generally in urban renewal schemes and in central area development. They can be effective but they soon deteriorate if they are not well maintained, kept free from weeds and regularly watered. Plant containers may add something to a small shopping precinct if proper arrangements can be made for them to be looked after.

APPENDIX A USE CLASSES ORDER, 1972

The order specifies 19 different use classes and these can be conveniently grouped as follows.

Class I	Shops but specifically excluding shops selling pets, cat meat, tripe and hot food and motor vehicles. It also excludes launderettes, cafés and restaurants
Class II	Office
Class III	Light industry
Class IV	General industry
Class V to IX	Special industries (usually those which give off smells, smoke, dust, etc.)
Class X	Warehouses
Class XI	Hotels and boarding houses
Classes XII to XIII	Schools, colleges and places of worship
Classes XIV to XVI	Hospitals, nursing homes, health centres and places of detention
Classes XVII to XIX	Public buildings and places of public assembly

The above summary is given to illustrate the principle, and detailed reference should be made to the Order in cases of doubt or when a material change of use is contemplated.

APPENDIX B CONTROL LEGISLATION

The following Acts of legislation all authorise the exercise of some form of control over development in their respective fields; these are in addition to those referred to in the text of this book.

Legislation	Field
Thermal Insulation (Industrial Buildings) Act, 1957, and Regulations, 1972	The insulation of roofs to certain factories.
Factories Act, 1961	Sanitary accommodation. Health and welfare.
Clean Air Act, 1956 and 1968	Height of chimneys to industrial premises. Treatment of offensive fumes from appliances. Types of installations.
Offices, Shops and Railway Premises Act, 1963	All matters as for Factories Acts.
Housing Act, 1957, 1961, 1969 and 1974	Housing standards and improvement of existing dwellings.
Highways Act, 1959, 1961 and 1971	Matters affecting streets and line of existing buildings on the street.
Public Health (Drainage of Trade Premises) Act 1937 and Public Health Act, 1961, Part V	Discharge of and treatment of trade effluents.
Control of Pollution Act, 1974, Parts II and III	Pollution of water. Control of noise.
Land Drainage Act, 1961	Provision for adequate land drainage and protection of existing drainage ways.
Explosives Act, 1875 and 1923	Use and storage of explosives.
Petroleum (Regulation) Act, 1928 and 1936	Use and storage of petroleum and similar substances.
Radio-active Substances Act, 1960	Use, storage and disposal of radio-active substances.
The Coal Act, 1938, Coal Mining (Subsidence) Act, 1957, and Open Cast Coal Act, 1958	Deal with effects of subsidence on buildings.

Water Act, 1945, and Water By-laws	Water supplies, etc.
Ancient Monuments Act, 1931, Historic Buildings and Ancient Monuments Act, 1953, and Field Monuments Act, 1972	Preservation and protection of ancient monuments.
Licensing Act, 1949, 1961 and 1964	Control of premises used for singing and dancing and for the sale of intoxicating liquor.
Rights of Way Act, 1959	Protection of public right of way.
The Public Utilities Street Works Act, 1950	Making good protection and reinstatement of road surfaces disturbed by underground services.
Food and Drugs Act, 1955	Controls standards of premises where food and/or drugs are stored and used.
The Fire Precautions Act, 1971	Means of escape in case of fire in designated buildings.
Health and Safety at Work etc. Act, 1974, Part I	Health, safety and welfare of employees.

APPENDIX C A DEVELOPER'S BRIEF

In order to set in motion the detailed planning and afterwards the building of a local authority housing estate, the chief housing officer or other appointed local government officer, having first obtained appropriate approvals and purchased the land, will normally prepare a brief. All the information available which will help in the detailed planning must be given in the brief or it must state where additional information can be obtained.

The brief should be made up of sections and these should be placed in an order as near as possible to the sequence of events they deal with. The make-up of the first section will be common to most briefs but thereafter they will vary according to the type of development proposed.

THE BRIEF

1. Originator and Date of Issue

2. Addressees

It is sometimes convenient to state early in the brief to whom the brief is addressed. In other cases it may be desirable to leave the information until a later stage in the brief and give it under the heading of 'Circulation'.

THE SITE

3. Location and Size of Site

This section will deal with the size and location and will define the exact boundaries on a plan of suitable scale.

4. Objective

At an early stage in the brief it is important to know at least in general terms the type of development and the time-scale proposed.

5. Restraints

Before any detailed plans can be worked out it is essential to know of all the restraints and other limiting factors which will affect the planning of the scheme. These should be given in groups for easy reference and the following sequence has been found satisfactory.

(a) Covenants and other restrictions which go with the land such as

    (i) restrictive covenants

    (ii) leasehold and freehold interest held by other parties with details of occupations and security of tenure

    (iii) easements and other rights over the land.

(b) Restraints imposed by planning legislation and other statutory controls. These may include

    (i) conservation orders

    (ii) listed buildings

    (iii) tree preservation orders

    (iv) vehicular and pedestrian access to site.

(c) Limitations resulting from the physical features of the site. These will generally be apparent on inspection of the site but there may be some things of which prior knowledge may save a lot of time later and this could include such things as

    (i) underground workings

    (ii) made-up land

    (iii) high water table.

Much of the above can be shown on a plan so that the building or land to which it refers can be quickly identified on the ground.

6. Existing Road Systems

Reference should be made to classification where appropriate and attention should be drawn to any private or unadopted roads.

7. Existing Services

These could also restrict a development. Most of the services can be moved or adapted to meet new requirements but difficulties may arise in respect of cost.

8. Transport Facilities

A note of existing transport facilities should include the location of public transport stops and terminals. If there is a railway station conveniently near, the direction of the entrance should be shown.

9. The Locality

Although this section is not in all briefs, there are cases where it would be helpful to give a general note of other types of development in the locality.

SITE DEVELOPMENT

10. Accommodation Required

A brief for a scheme of residential development for a housing authority would contain very detailed requirements but in order to keep this section to a

manageable size, section 14 deals with internal space distribution, restricting this section to such matter as
(a) the approximate number and types of dwellings
(b) special requirements in respect of old people, disabled people, etc.
Details of internal circulation and space are dealt with in section 14.

11. Standards

This section deals with standards to be aimed at and should include

(a) residential density
(b) daylighting standards
(c) cost levels, including reference to the prevailing cost yardstick
(d) Parker Morris or other standards
(e) the National House Building Registration Council.

12. Form of Construction

The form of construction should only be stated in general terms in this section

(a) conventional, say, loadbearing brickwork
(b) prefabrication
(c) system built
(d) other methods.

13. Layout and Environmental Requirements

This section deals generally with the layout of the site and will state any particular requirements such as

(a) Radburn or vehicular cul-de-sac
(b) design guide
(c) housing and garage courts
(d) size and treatment of frontage
(e) size of individual gardens
(f) other enclosures
(g) open amenity space
(h) sitting areas
(i) children's play areas
(j) garages and parking standards for both residents and visitors
(k) proportion of garages, car ports and hard standings
(l) shopping provision within the scheme
(m) community buildings
(n) landscaping (including method of upkeep)
(o) refuse collection
(p) communal TV and radio aerials
(q) laundry and drying provision.

164

14. Internal Designs

The actual designs and individual dwellings are not dealt with in this section but it contains a statement of requirements under the following subheadings

(a) space standards
(b) circulation and space distribution
(c) fittings and equipment to be supplied
(d) electrical fittings for light and power
(e) fuel and type of stove for cooking and space heating
(f) storage of fuel
(g) standards for space heating.

15. Further Information

This section will state the names, addresses and telephone numbers of persons or organisations where further information can be obtained.

16. Administrative Arrangements

All the procedural and other details required and connected with making a bid or taking up an invitation contained in the brief are given in this section. It should state in detail any time-scale laid down for submission and acceptance and whether there is any provision for an extension of time. It should contain protective clauses concerned with the acceptance of tenders.

17. Circulation

Some briefs are given wide circulation and others are restricted to certain addressees. There may also be a degree of confidentiality. In some cases each copy is given a number and allocated to one person or firm by name. Repetition should be avoided if the matter is dealt with in section 2.

18. The Appendixes

In this section there will generally be maps and plans but in certain circumstances there may also be calculation and other details which although important are better kept out of the main script.

APPENDIX D OUTLINE OF PROCEDURE FROM THE STREET WORKS CODE, 1892

Prior to the reorganisation of local government there were two different procedures for the adoption of streets and they were known by the following numbers: 1885 and 1892. The significance of the numbers is that they are the dates when the respective codes were enacted. The Street Works Code, 1892, is now the only one in operation, and is contained in the Highways Act, 1959.

Where a street is not sewered, levelled, paved, metalled, flagged, channelled and made good to the satisfaction of the highway authority they may resolve to do any or all of these works with the cost to be apportioned on the premises abutting on the street. (There are certain exceptions which are given later.) The following steps will be taken. The surveyor will prepare

A   Specification with plans and section
B   Estimate of cost (expenses)
C   Provisional apportionment of the cost to be charged on the respective premises. It should give the name of the owner and state if the apportionment is based on frontage, giving the measurement used. If any other considerations have been taken into account they should also be stated

The highway authority approves by resolution, which has to be published for 2 successive weeks in a local paper and posted in or near the street to which it relates for 3 weeks, and copies served on the owners of the premises liable. Copies of the specification and other documents are made available for open inspection.

Objections may be made on a number of grounds, including

(1) that the alleged street is not a street within the meaning of the Act
(2) that there was or is informality in some stage of the proceedings
(3) that the proposed work is insufficient or unreasonable
(4) that the estimated expenses are excessive
(5) that certain premises should be excluded from or inserted in the provisional apportionment
(6) that the provisional apportionment is incorrect.

Objections are heard and determined in the magistrates' court.

The highway authority may depart from an apportionment strictly related to frontage and consider

(1) the benefit to be derived by any premises
(2) the value of any work already done by owner or occupier.

They may also include in the apportionment premises with access to the street without owning a frontage.

After completion of the work the surveyor will divide the expenses in accordance with the original apportionment and serve notice on the owners who may object within 1 month on any of the following grounds

(1) unreasonable departure from the specification, etc.
(2) the actual expenses exceeded the estimate by over 15 per cent without sufficient reason
(3) that the final apportionment was incorrect.

There are also grounds when the frontagers can require the highway authority to make up and adopt a private street

(1) Those frontagers who wish to exercise these powers must be made up of a majority of frontagers on their length of the street; or the total of their frontages must be more than half the total length of frontages on both sides of the street.
(2) Where the length of the street is more than 100 yards.
(3) Where payment has been made or secured by at least one frontager under the Advanced Payments Code.

A majority (in rateable value) can compel the highway authority to take over a street in which all the works have been satisfactorily carried out.

APPENDIX E   A SUMMARY OF THE ADVANCE PAYMENTS CODE (HIGHWAYS ACT, 1959)

This code provides that before buildings are erected fronting on to private streets (this includes proposed streets) the sum likely to be needed for street works is deposited with the highway authority or security is given for it.

A notice must be served on the owner within 1 month of the plans being submitted for Building Regulations approval. The owner may appeal to the Minister also within one month.

The authority must pay interest but this remains on deposit with the principal until the street works are carried out.

If, when the work is carried out, the costs are less than the amount on deposit a refund will be made to the owner. Conversely, the highway authority has the power to claim the difference if the cost is higher.

When the street works are carried out the highway authority may, by a notice fixed in the street, declare it to be 'maintainable at public expense' and it will become so unless the majority (in numbers) object. The highway authority may apply to the magistrates' court to overrule the objections.

There are quite a number of cases where the above provisions do not apply. They are too numerous to reproduce in detail and are too complicated to reduce to a summary. They are mainly concerned with

(1) cases where other provisions are made
(2) the length of the street
(3) the time likely to elapse before it is fully developed
(4) cases where the street does not run into a made-up road
(5) cases where part of the street is occupied by industrial premises.

The magistrates' court hears these objections.

Certain frontagers are exempt and these include churches, their proportion being paid by the highway authority. Railway and canal authorities are also exempt, provided they have no direct communication with the street, but in these cases the cost is borne by the other owners.

There are provisions for a re-apportionment if at any future time direct communication is made, and for payments to be refunded to owners.

APPENDIX F ARNDALE CHRONOLOGY

The construction of the Luton Arndale Centre began in December, 1969.

The project has been described by the developers as one of the most complex redevelopment schemes ever undertaken outside London.

To make way for the centre, the Council had to acquire a total of some 600 business interests, ranging from leading stores to small, individual cottages. The land required for the redevelopment scheme was acquired either by agreement or by means of three major Compulsory Purchase Orders. Total cost of acquisition was approximately £6 million. The relocation of displaced traders who wished to go into the centre has already taken place in the vast majority of cases. The remainder who wish to do so will move into later phases.

The planning of the demolition and construction process had to be meticulous. Until new accommodation became available for a displaced trader and open to the public, the existing premises could not be demolished. Each new section of the centre has to be commercially viable. Disruption of the commercial activity of the town centre must be kept to a minimum. Shortage of storage space for materials presents added difficulties to the job of construction.

The choice of structural steel as the main constructional material has helped to solve a number of problems. A steel framework was the first form of modern, industrialised building and is one of the most effective, the crucial factors being the ease and speed of assembly.

These are the key dates in the development of the Luton Arndale Centre in the first 10 years.

November	1963	Comprehensive Development Area designation approved
February	1965	Compulsory Purchase Order No. 1 confirmed
March	1965	Developers invited to submit schemes
March	1966	Short list of three schemes selected
December	1966	Scheme and offer selected
November	1967	Compulsory Purchase Order No. 2 confirmed
April	1968	Scheme and financial offer accepted by Council
December	1969	Construction commenced
February	1970	Compulsory Purchase Order No. 3 confirmed
August	1971	Library Car Park opened
January	1972	Market, and Market Car Park opened, plus the Student Prince public house and nine shops
November	1972	A further three major stores and 32 shops opened, plus the national headquarters of the Local Government Training Board. Two more department stores now in use
May	1973	Opening of Strathmore Hotel

APPENDIX G   TREES IN LANDSCAPING

Common name	Spread (metres)	Class	Growing conditions	Foliage
Large trees 12.5 m upwards				
Oak	15	broad		mid-green
Beech	15	broad		mid-green (pale green in spring)
Copper beech	15	broad		copper
Silver maple	15	round	town	pale green (yellow in autumn)
Horse chestnut	22	round		dark green white flowers
Sweet chestnut	18	round		dark green
Plane	22	round		mid-green (yellow in autumn)
English elm	15	square	town	dark green (yellow in autumn)
Norway maple	22	square	town	dark green (golden in autumn)
Wych-elm	25	square	coastal	dark green (yellow in autumn)
Alder	8	tapering	waterside	dark green
White poplar	20	tapering	waterside	white
Bat willow	12	tapering	waterside	blue gray
Poplar	10	conical		dark green
False acacia	15	conical	town	pale green
Lawson's cypress	5	columnar		dark green (evergreen)
Lombardy poplar	5	columnar		dark green
Medium trees 7.5 m upwards				
Whitebeam	10	round		gray/green
Red horse-chestnut	10	round		dark green (red flowers)
Holly	6	conical	town	dark green (evergreen)
Mountain ash	9	conical		dark green (berries in autumn)
Almond	7	round	town	dark green (pink flowers in spring)
Winter cherry	6	round		mid-green (white flowers in autumn)
Hawthorn	6	round		dark green (red or white flowers)
Yew	3	columnar		dark green
Juniper	3	columnar		dark green (evergreen)

INDEX